The Philosophy of Physics

To poor little Nora (who always gets left out ...)

The Philosophy of Physics

DEAN RICKLES

polity

The right of Dean Rickles to be identified as Author of this Work has been asserted in accordance with the UK Copyright, Designs and Patents Act 1988.

First published in 2016 by Polity Press

Polity Press
65 Bridge Street
Cambridge CB2 1UR, UK

Polity Press
350 Main Street
Malden, MA 02148, USA

ISBN-13: 978-0-7456-6981-6
ISBN-13: 978-0-7456-6982-3(pb)

A catalogue record for this book is available from the British Library.

Library of Congress Cataloging-in-Publication Data

Names: Rickles, Dean.
Title: The philosophy of physics / Dean Rickles.
Description: Cambridge : Polity Press, 2016. | Includes bibliographical
    references and index.
Identifiers: LCCN 2015043550 (print) | LCCN 2015047675 (ebook) | ISBN
    9780745669816 (hardback) | ISBN 9780745669823 (pbk.) | ISBN 9781509509393
    (Mobi) | ISBN 9781509509409 (Epub)
Subjects: LCSH: Physics--Philosophy.
Classification: LCC QC6 .R4635 2016 (print) | LCC QC6 (ebook) | DDC
    530.01--dc23
LC record available at http://lccn.loc.gov/2015043550

Typeset in 9.5/13 Utopia by
Servis Filmsetting Ltd, Stockport, Cheshire

For further information on Polity, visit our website: politybooks.com

Contents

Preface

Philosophical thinking brings with it a heathy dose of skepticism about many things that are often taken for granted. This is generally a good thing and can lead to greater clarity. It can also lead to ideas that look downright bizarre to the general population: the idea that you might be a brain-in-a-vat; the idea that the universe could have come into being five seconds ago; the idea that you are really a spacetime worm stretching from your birth to your death!

In this book we apply the principles of philosophy to theories of physics – it is important to emphasize that we are dealing with *theories* of physics in philosophy of physics: those are the entities that link us to the world and are that which we must interpret. Again, this involves clarity through skepticism, but also leads to views that strike many as odd: the idea that it takes a mind to 'collapse a quantum state;' the idea that there are many worlds (one for each choice in quantum mechanical experiment); the idea that the future can causally influence the present. The good news is that these latter examples are grounded in our best physical theories: odd or not, they seem to offer possible interpretations of these theories (i.e. they are 'ways the world could be given the truth of the theory'). Modern physics can even throw some light on what might be considered 'purely philosophical' issues, such as 'what is the relationship between an object and its properties?', 'can two objects share exactly the same properties?', 'is the future open or fixed?', 'does time flow?', and 'are you really a four-dimensional spacetime worm?'

There are several books on philosophy of physics that do a very good job at introducing the reader to the basic issues, with varying levels of difficulty. The present book aims to provide a snapshot of the central topics, methods, and problems of modern philosophy of physics in a very elementary manner. The audience for this book is the absolute beginner, albeit one with a modicum of mathematical ability (or an ability to at least not glaze over at the mere sight of some mathematical formulae). But it nonetheless aims to be a *complete* course, in that it covers all the main areas (classical and quantum, relativistic and non-relativistic, statistical and non-statistical) and provides supplementary readings across a range of skill levels, so indicating the work that needs to be accomplished to reach research-level philosophy of physics.

The book is written for both early stage philosophers *and* physicists:

- Physicists who want more than rules and algorithms for churning out numbers to compare with experiment, and want to know 'what it all means.'
- Philosophers who want more than watered-down, quasi-journalistic physics, and who are not satisfied with philosophical discussions of the nature of the physical world that do not engage with physics.

I'm not going to tell you to shut up and calculate; but I'm not going to tell you *not* to calculate either! A little bit of computational skill is vital in good philosophy of physics, though one could probably be a good physicist without having a philosophical bone in one's body. I would argue, however, that one could not be a *great* physicist without a good head for philosophical thinking. Despite the necessity of having a good grasp of the mathematical details of physical theories in order to be a proficient philosopher of physics, this book makes do with the bare minimum: it is a stepping stone to the many books of a rather more mathematically involved nature. Where matters get a little technical (or where there are interesting diversions, historical or otherwise), I have relegated these to endnotes – these often contain suggestions for interesting, usually more advanced readings.

The book begins, in Chapter 1, with some general considerations about philosophy of physics itself, as a discipline. Central to this is the idea that philosophy of physics concerns itself primarily with *interpreting* the representations of physical systems that can be found in physics (usually the best available physics, couched in mathematical representations). We also consider the question of why these mathematical representations seem to be so good at gripping onto the world.

Chapter 2 introduces some basic concepts from physics: the states, observables, and dynamics that form the bricks and mortar of the world-pictures (or 'ontologies') according to our theories. This machinery is then used to introduce symmetries in physics in Chapter 3, in which we also begin to see how philosophical issues emerge from symmetry – symmetry will play a central role in the chapters that follow.

Chapters 4 to 7 apply all of this foregoing discussion to specific examples, starting with spacetime theories in Chapters 4 and 5 (including the theories of relativity), then statistical physics (in Chapter 6), and quantum mechanics (Chapter 7). Chapter 8 provides 'tasters' of seven more cutting edge issues in philosophy of physics, which might provide more scope for future research projects for budding philosophers of physics.

Each chapter includes a handful of further readings, organized according to difficulty: 'fun,' 'serious,' and 'connoisseur' level. This book should be supplemented with at least the fun readings (or suitable extracts) in order to provide a rounded picture. Together with these, and perhaps several of the readings from the endnotes, this would provide material enough for a semester-long course in introductory philosophy of physics.

I should perhaps end with a brief note on the endnotes in this book: there are lots of them! P. G. Wodehouse mercilessly slammed footnotes in his autobiography *Over Seventy: An Autobiography with Digressions* (Herbert Jenkins, 1957):

> When I read a book I am like someone strolling across a level lawn thinking how jolly it all is, and when I am suddenly confronted with a [1] or a [2] it is as though I had stepped on the teeth of a rake and had the handle spring up and hit me on the bridge of the nose. I stop dead and my eyes flicker and swivel. I tell myself that this time I will not be fooled into looking at the beastly thing, but I always am, and it nearly always maddens me by beginning with the word 'see.'

However, endnotes, while ameliorating some of Wodehouse's complaints, also face his wrath:

> Slightly, but not much, better than the footnotes which jerk your eye to the bottom of the page are those which are lumped together somewhere in the back of the book. These allow of continuous reading, or at any rate are supposed to, but it is only a man of iron will who, coming on a [b] or a [7], can keep from dropping everything and bounding off after it like a basset hound after a basset.

Some explanation is therefore in order for filling a book with almost ninety endnotes. These notes serve two functions:

1   to cope with the inevitable difference in skill sets and diverse backgrounds of the likely readers of this book: some of you will be at ease with philosophical concepts, but perhaps confused by certain aspects of mathematics and physics; and some of you will have the opposite problem. Endnotes provide some opportunity for filling out such concepts.
2   to enable readers' specific special interests to be attended to (mostly via suggestions for further readings) in an unobtrusive way (Wodehouse's remarks notwithstanding): some of you would be keen to know where they can learn more about conventionalist principles in science, for example, while others find that incredibly boring and will want to focus on the more mind-boggling, metaphysical issues, or some more technical issues.

A reader with no inclination to bounce around the book can rest assured that (again, Wodehouse's remarks notwithstanding) the entire book can be read without ever turning to the back, or perhaps browsing through them separately at their leisure.

Sydney, October, 2015   *DPR*

Acknowledgments

This work would not have been possible without the very generous financial support of the Australian Research Council (through an Australian Research Council Future Fellowship: FT130100466). I also thank the various students that took my philosophy of physics seminar, HPSC4101, at the University of Sydney, that allowed me to try out many of the ideas in this book. Especially: Ariella Adler, Jessica Bloom, Prashant Kumar, Benjamin Pope, and Ann Thresher. Thanks also to Alex Zachary (for her laser-typo eyes) and Lou James (for pretending to find my jokes amusing). I'm deeply grateful to Feraz Azhar for his very detailed comments on an earlier draft, and likewise to Jason Grossman for many helpful suggestions. Pascal Porcheron and Ellen MacDonald-Kramer also deserve thanks for sticking with this book as it glided through deadline after deadline. Sorry if I missed anyone!

And, as always, the customary doffing of the cap to the family for putting up with the many late nights during which this book was completed! Gaia deserves a special thanks for the 'high-entropy configuration' (= her bedroom) photograph – admittedly it wasn't so hard for her to engineer, but still. . .

1 Interpreting Physical Theories

In this chapter we get to grips with what philosophy of physics is all about and what kinds of questions it deals with. We also introduce some basic general concepts and terminology from philosophy: ontology, epistemology, etc. And also introduce such essential philosophy of science concepts as 'theory,' 'model' and so on – the concepts on which our later discussions will be based (the tools of the trade, so to speak). A key point that will be emphasized here is that there is often a difficulty in understanding how some (empirically) successful theory (formulated in 'the language of mathematics') can map on to physical reality – there is an additional question discussed of *how* mathematics can perform its feat of allowing the formulation of precise, successful physical laws. There is often, for example, a multiplicity of possible 'ways the world could be' according to the mathematical structure, while still preserving the theory's empirical success. We indicate that symmetries often lie at the root of (the most interesting of) these situations – a fact that will form the basis of much of this book.

1.1 Does the World Need Philosophers of Physics?

Does the world really have any need for philosophers of physics and the odd trade they ply? If we were to put them all on a spacecraft, in the style of the *Hitchhiker's Guide to the Galaxy*, say, on the Golgafrincham Ark Fleet's 'Ship B,' occupied by telephone sanitizers, public relations executives, advertising account executives, and other such 'worthy' tradespeople, would the human race be worse off or all the better for it?

Of course "need" can mean many things, each depending on the purpose or use. If we believe that there is a need to think especially deeply about physical theories, about why they work so well, and what they can tell us about the nature of reality: then there *is* a need for philosophers of physics. If we mean is there a *practical* use for philosophy of physics, then it is less clear that they have anything to contribute to the world, though that is not a clear-cut matter. But the main problem we face in answering this question is knowing where to draw the line between physics and philosophy of physics. If we were to end up throwing David Bohm, Niels Bohr, Albert Einstein, Werner Heisenberg, Erwin Schrödinger, and Hermann Weyl onto Ship B then you would agree, I hope, that the world would be all the worse for it. Yet each of these physicists was a 'philosopher-physicist': they even wrote books on the philosophy of physics. In days gone past they would have been called 'natural philosophers,' like Newton. I think *natural philosophy* is still a very useful term to use to describe one who studies the natural sciences philosophically, and if I had my way it would be back in operation to describe philosophers of physics and philosopher-physicists. I can think of several living (as of 2015, and long may they continue) physicists that fit the traditional mold of natural philosopher, among them Julian Barbour, Rudolph Haag, Roger Penrose, Carlo Rovelli, Lee Smolin, Max Tegmark, Gerard't Hooft, and Dieter Zeh. What characterizes them is that they think deeply about the *foundations* of their subject, and especially about the nature of space, time, and matter: our primary subject matter (or rather, the subject matter *of* our subject matter: *theories* of space, time, and matter).

And yet the philosophy of physics is often frowned upon by physicists. When physicists lapse into discussing philosophy it is seen to be just that: a lapse. Richard Feynman is famously reported to have said that "Philosophy of science is about as useful to scientists as ornithology is to birds" – there is a fairly sizeable catalogue of Feynman's anti-philosophy quotes to draw from and I think much of today's anti-philosophical spirit has a lot to do with Feynman worship (though there are many worse people to worship. . .). Again, as above, this might well be true, of course, depending on how one interprets 'useful.' 'Use' is just like 'need' again. When it comes to computing values of physical quantities to be compared with experiment, admittedly philosophers might not be of much use. They won't be too upset that they aren't of use in this sense: it's not what they live for. But if one is facing some problem in the foundations of physics, then the more generalist approach of a philosopher (or a philosophical *approach*) might be of some use after all – likewise, ornithology might well be useful to birds, in terms of conservation of a species for example!

Need and *usefulness* need to be more carefully circumscribed. One can easily make a case for a need for a philosophical approach to physics, and for the usefulness of such an approach. Given this, one can make a case for having a specific discipline, a dedicated community of scholars, devoted to such an approach, putting aside as secondary those core aspects of physics itself. Hence, becoming a philosopher of physics rather than a physicist involves a trade-off: you put the computing-intensive aspects aside in favor of the critical, interpretive aspects. Some people can do both, but they are the exception.

A philosophical *approach* to physics will direct attention to aspects of physics that are usually deemed sacred by the average practicing physicist. This can lead to advances by opening up new lines of enquiry, suggesting hypotheses that would be unthinkable at the everyday level of physics. In this book we will meet several examples in which this has occurred, mostly based on the work of philosopher-physicists. One often finds that thought experiments (or *gedanken* experiments) lie at the root of the really major advances in physics, the revolutions. Such thought experiments usually probe some foundational assumption, concerning space, time, locality, causality, determinism, matter, force, and so on. This kind of approach amounts to philosophical thinking (in the sense of conceptual analysis).

In this sense, then, philosophy is a fundamental part of the development of physics. One can't really do physics without making certain assumptions, however minimal, about how the theories one is using map onto reality (even if it is only mapping onto *observable* reality, as some believe [anti-realists, such as instrumentalists or constructive empiricists, who give unobservables the 'silent treatment'], though even this is not so benign a claim as they think).[1] One needs to take a stance on what the components of the theories refer to, what they are *about* (this is the meaning of the philosophical term 'ontology': what there is).

This is one of the primary functions of philosophers of physics: interpretation of physical theories (on which, see §1.3 below). Mapping between theory and world. It is almost never a trivial matter; especially since physical theories often make use of all sorts of idealizations, approximations, and indirect methods of representing their 'target' systems. A key assumption underlying philosophy of physics, then, is that the job of physics is to say something about the structure of reality, about what the universe is *really* like: what objects there are, what properties they have, how they behave, how they relate to one another, and so on. Much of this book will be devoted to specific examples of this type and will highlight the ways in which interpretive controversies emerge. The interpretive controversies point to the existence of 'epistemological' considerations: what we can know. If there are multiple possible interpretations of some theory, then it seems we are limited in what a theory can tell us about the world. We will meet this in a very stark way in §5.1, in which an argument is presented that claims to show that our choice of world-geometry is largely conventional (that is, there is no fact of the matter that can decide, so we use other considerations to choose). We consider what interpretation amounts to in §1.3. First, let's take a look at a historical feature of philosophy of physics: the issue of its relative recency.

1.2 Why No Ancient Philosophy of Physics?

Philosophy of physics is a fairly modern discipline, emerging from various specializations that occurred at the close of the nineteenth century. Why didn't philosophy of physics exist earlier? Why wasn't there such a thing in the days of Democritus and Parmenides? We are happy to still discuss Newtonian mechanics as philosophers of physics, but it seems that stretching further back doesn't quite work in the same way (even with Plato and Aristotle): why is this? One important factor is that the pre-Socratics, and those working before the Medieval period, relied on *naked eye* observations. There were no amplifications of vision as there have been since Galileo modified the looking glass (until then a simple gypsy toy) into a telescope for scientific usage. There was, in fact, no real experimental method. Some kinds of experiments could no doubt be said to have occurred, but this was not viewed as the royal road to worldly knowledge as it is today. This meant that theoretical structures were far more heavily based on what was delivered through the senses. However, as we will see in a moment, this wasn't always the case, and unobservables invoked to explain observables can be found in theories in the earliest fragments of writing.

Modern philosophy of physics would be unimaginable without theories that can be put into a fairly standard mathematical form. For example, spacetime theories are presented through 'models' of the form $\langle \mathcal{M}, \mathcal{S}^i \rangle$, consisting of some basic set of point-elements $\mathcal{M}$ (a set of points with a certain size or cardinality) on which is imposed various levels of additional structure $\mathcal{S}^i$ that let one talk about, e.g. the dimension of the space, the nearness of points, the distance of points, volumes, parallelism, and so on. This then provides the basic object whose mapping onto the world (or *a* world) we must, wearing our philosophy of physics hats, consider – that is, we must provide an interpretation. This will involve setting up a correspondence between mathematical entities, ⟨the set of mathematical points and mathematical relationships between such points⟩, and ⟨things and their properties in the world⟩.

This provides a second factor for the relative recency of philosophy of physics: earlier work was less mathematical, or entirely non-mathematical. The development of calculus in particular was pivotal in the development of a physics that was able to make precise predictions and allow thinking in terms of present (initial) conditions *generating* future states. Also, more conceptually, the notion of a 'law of nature' (understood as an expression of *invariance* or *constancy*) was fairly slow to emerge, and can be seen to receive its enunciation with Galileo – though, even here, it is expressed in the form of a dialogue rather than in formal terminology.[2]

In fact, inasmuch as there were *theories* proposed by Democritus and Parmenides, there is something that might very loosely be called 'philosophy of ancient physics' (i.e. rather than ancient philosophy of physics). Those theories tended to focus on cosmological issues and were largely *a priori* (that is, based on reason and logic rather than experience). What we have from these philosophers are mere fragments. In the case of Parmenides we only have a fragment of a poem (160 lines of a supposed

800-line work). However, we can find strikingly similar dichotomies to those found in modern theories: discrete versus continuous space; infinite versus finite space; real time versus illusory time; eternal time versus beginning of time; plenum versus void; many worlds versus one world; and so on. There are also discussions of motion and its relationship with space and time, along with the ramifications a denial of the former has on the latter, and vice versa. Let us mention some of these views, since they provide a nice route into some of the more modern debates we discuss later.

Though I downplayed the mathematical nature of early theories, Zeno can with some justification be said to have considered a mathematical account of space, time, and motion, if only to dismiss its applicability to the real world, on pain of generating paradoxes. Take, for example, the so-called 'paradox of plurality.' Here we are asked to consider an extended object: a 5 cm long line, say. We can envisage splitting this into two pieces, and then splitting those two pieces each into two more pieces, and so on. How will this process end if space is infinitely divisible? Zeno suggests two possibilities: (1) with points of some finite (though minuscule) size, or (2) with points of zero size. But neither is acceptable, says Zeno: if the points have some size, then given we have made infinite divisions, there will be infinitely many, which will certainly not generate a 5 cm line when aggregated. But if the points have zero size, then not even an infinity of them will aggregate to form a 5 cm line. Or consider 'the arrow paradox.' Here the punchline is that an arrow in flight can never be in motion at any instant, since it will be at rest at an instant; if it weren't, the instant must have parts, and so not be indivisible. And yet it somehow gets from one instant to the next. In some ways this is a motion-based analogue of the plurality paradox: how can we add up lots of stationary pieces to get an extended motion? The stimulating effect on future physics (and mathematics) cannot be overestimated.[3]

If a mathematical account of space, time, and motion (in terms of continuous, infinitely divisible entities and processes) is to be carried out, then Zeno's challenges would have to be met. If they could not be met, then at the very least there would be a gap between model and reality. Atomism was one early response to the challenge: if space (and perhaps time) could not be infinitely divided, then Zeno's division paradoxes don't hold water. In any case, they shift attention to the possible structure of space and time: what kind of things they are. And motion is implicated in this structural exploration.

We can also fit earlier theories, such as Aristotle or Plato's cosmology, into the mold of the model-based approach mentioned above. In these cosmologies there is a preferred location in the universe corresponding to the Earth's position in space, which is assumed to be at absolute rest (which makes intuitive sense, of course, since we don't *seem* to be moving

Fig. 1.1 A woodcut by Camille Flammarion showing Archytas' thought experiment in action, breaking through Aristotle's closed universe with his stick and his hand, to reveal more space.
[Source: C. Flammarion, *L'Atmosphere: Météorologie Populaire* (Paris, 1888: p. 163)].

when we ourselves remain at rest relative to the Earth). The universe is, in this scheme, spherical, with the Earth at the dead center, stationary and not rotating, and the Moon, Mercury, Venus, the Sun, Mars, Jupiter, and Saturn occupying (in this order outwards from the Earth) concentric spheres – beyond this lies 'the firmament,' an outmost layer of fixed stars providing an 'edge' to the universe (an idea mocked by Archytas who asked what would happen if we poke a stick or a hand outwards from this layer – see fig. 1.1). This allows for talk of absolute distances (and absolute notions of 'up' and 'down') by using the Earth's perspective as the origin. Of course, the Copernican revolution would dislodge this special perspective by dislodging the specialness of the location of the Earth and its relationship to the other planets and stars.

In the Aristotelian universe there are five elements (Aether, Fire, Air, Water, and Earth) that have 'natural places' that determine their behaviors (natural motions) relative to the Earth's frame. This is a *teleological* world in which things are guided by where they *should* be: fire wants to rise; earth wants to fall. This is a law of sorts, and it is based to some extent on observations. But it certainly doesn't have same level of rigor of modern physical laws, nor does it strike one as particularly explanatory, nor does it seem capable of generating very interesting predictions. Rather, the fact that, e.g. the stars are found to move in a circular fashion, is made a necessary part of the world: this is what stars *must* do, and such things constrained to move in circles should be given a special name (this circular motion is the essential quality of *aether* from which the heavenly bodies are made). This circularity property was a key part of Plato's *Timaeus*, in which his cosmology was laid out. Later, in his *Almagest*, Ptolemy converted this idea (of circular, uniform planetary motion) into a mathematical model (based on nothing but circular regular motions) capable of delivering predictions of our apparently rather messy, irregular world. 'Model' in this sense is meant in terms of an approximate representation, *not* an exact one-to-one correspondence with 'the way things are.' Philosophers since Plato have spoken of this kind of modeling as 'saving the phenomena': the recovery (by means of a theory or model) of the way the world *appears to us*.

But, more importantly, there is a detailed discussion of the notion of space in Aristotle and Plato. This was identified with 'place' in their schemes. Aristotle's account is more detailed. His view is 'plenistic,' meaning that space (the universe) is viewed as *always full* (with no genuinely empty spaces): if some portion of space is not occupied by water or matter it is occupied by air, each displacing one or the other as they are moved. As mentioned above, this was combined with a theory of 'natural place' whereby each kind of thing is transported to its own specific category of place (hence, the teleological notion of a 'final' or 'future cause' in this case). The notion of a natural motion (in which a body finds its 'proper place') and a 'forced (or "violent") motion' (in which a body is taken off its proper course by some intervention) is the origin of the modern concepts of *kinematics* and *dynamics* respectively, which we will meet again.

In his *Timaeus*, Plato also defends the view that the world is a plenum: matter and space are, in fact, identified. Aristotle, however, introduces *place* as a kind of entity independent from matter, since while matter (and the form the matter takes) are essential to a body, place is not: different objects can occupy the same space, and the same object can occupy different places. Place itself is defined in terms of objects, however. It is the two-dimensional boundary of the body it contains, which it is in contact with. It is presented by Aristotle (in his *Physics, Book IV*) using the analogy of a (motionless) container, in fact. This metaphor persists into the modern debate between so-called 'substantivalists' (who view space as a real container) and 'relationists' (who view space as nothing over and above the relations between things) – see §4.1. However, Aristotle doesn't sit completely comfortably in either camp, though he is usually aligned with relationism since places are defined by the boundaries of objects: no objects, no boundaries, no places.

Time was, in Aristotle's picture, similarly related to physical entities, in this case changes (of places or properties of things): no change = no time. The *measure* of time was linked, then, to (uniform) motion – though it is not *identified* with motion and change: how could it be? (1) motion is attached to the thing moving, while time stretches out everywhere, and (2) motions vary in speed, but time does not. Indeed, Aristotle speaks of time's being a 'measure of motion' in the same way as the cubit measures lengths. In the case of time the unit (analogous to the cubit) is provided by the revolution of the outermost sphere of the heavens. But Aristotle can be found intellectually struggling with the status of time, as with space: are they 'real' or not (i.e. in the world or in the mind)? If they are real, how closely linked are they to their measures (cubits, revolutions, and the like)? For Aristotle, motion is ontologically (and logically) prior to time: the latter is defined in terms of the former. Such probing is at the heart of philosophy of physics, and the same questions (and often similar responses) arise again and again over the two millennia connecting Aristotle with us.

1.3 The Interpretation Game

If the primary task of the physicist is to construct models and theories of the world, the primary task of a philosopher of physics is to *interpret* these products of physics, be they theories, models, simulations, or experiments (including so-called 'thought experiments'). That is, suppose that we believe one of these constructions: what must the world be like to ground the belief? We needn't trouble ourselves with questions of truth and realism; though connected, that's strictly a separate matter. We can think about 'model worlds' (possible worlds) that ground the belief instead. Our own, actual world might be among this class of possible worlds, but it is not necessary in order to provide an interpretation (and an associated ontological picture).

In this book we shall focus on theories, models, and thought experiments and the possible relationship they bear to physical reality. Understanding what a theory or model says about reality is to offer an interpretation. As mentioned above, it involves 'mapping the theory to the world.' Usually it is no simple matter, and many possibilities are available. Hence we often face a *problem of interpretation* (for some theory or model) in that there can be many interpretations compatible both with the same theory or model and with what we observe with the naked senses or experiment. Of course, this demands a thorough discussion of what we mean by 'theory' and 'model' (the targets of our interpretations) – we gave a very brief description above, but the issue forms an entire project in philosophy of science.[4]

One can understand this situation as somewhat similar to the interpretation of a Rorschach test in psychoanalysis (see fig. 1.2). Here a variety

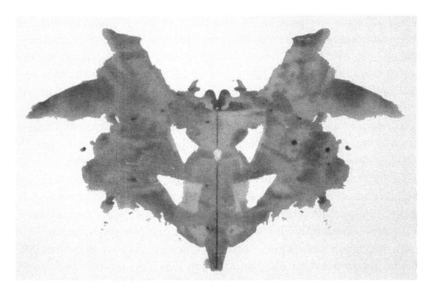

Fig. 1.2   A Rorschach inkblot test used to test for various personality disorders (originally a diagnostic tool for schizophrenia) and uncovering past, repressed traumas. What do you see? Most people see a bat in this example. But a variety of 'interpretations' are possible. Models and theories in science tend to be similarly multiply interpretable. [Image in the public domain]

of inkblots are shown to a subject who is then asked to comment on what he sees. The inkblots look random, but are (supposedly) well-chosen and presented in a sequence of ten ever more complex (and [supposedly] more revealing) blots. The subject will interpret the print in such a way as, so the idea goes, to give some clues as to the nature of their personality, including any disorders or past traumas they may have suffered and buried in their subconscious. If someone sees a fluffy bunny in the image, then we are led to think that they will not be an axe murderer, though they may be a little soft in the head. However, if a scene of human sacrifice is seen in the same image by someone else, then it's time to call in the men in white coats!

What we are faced with in physics is not inkblots but mathematical structures of some kind. Just as with the Rorschach test different subjects will interpret the self-same blot in many and varied ways, so one and the same mathematical formalism can (very often) be understood in many different ways. But this isn't a free for all: the ways one can interpret the formalism are highly constrained by the world (experience and experimental evidence) and by logical and mathematical consistency. However, even these tight constraints leave much elbow room, resulting in world-pictures entirely at odds with one another in all but empirical matters.

Given this feature, we see that 'interpretation' is a well-chosen term: in art we speak of "interpreting the painting" (usually when it is an abstract work). In music we speak of the performer as "interpreting the works of the great masters." Implicit in this is the idea of a *multiplicity* of interpretive options. Also related to the ordinary-language sense of interpretation is that there must be some 'closeness' to the painting or musical score. You can't very well give an interpretation of a Beethoven sonata without actually rendering sounds that are 'isomorphic' (that is, in one-to-one correspondence, or thereabouts) with the score (even if there are some wrong notes so that the isomorphism is partial, rather than a perfect correspondence). Different interpretations of a score will differ in some ways, but will have the same basic structure as supplied by the score, which supplies a kind of musical syntax akin to a mathematical formalism of a physical theory. Interpretations of physical theories can (must) be the same in some ways, while differing in other ways, just like musical interpretations.

Interpretation is also closely linked to ontology: to interpret is often just to provide an ontology. Bas van Fraassen describes the link as follows: "The question of interpretation [asks:] what would it be like for this theory to be true, and how could the world possibly be the way this theory says it is?" ([52], p. 242). The interpreter will then answer by specifying the class of possible worlds that make the theory true in the sense of satisfying the basic theoretical postulates (laws or axioms) – the musical performance analogy would state that interpretations of a piece of music (musical possible worlds) are those that satisfy the score. What we end up with, then, is a set of possible worlds that make the theory true; or, a set of possible worlds *according to the theory* (or *at* which the theory is true).

There are, then, two parts to an interpretation: a *syntactic* part (in which the formal structures and central axioms are laid out) and a *semantic* part providing the formal structures with 'meaning' (and in which the 'possible worlds' are specified, where these worlds are taken to be 'models' of the syntactic part). We would also have to consider some kind of 'relevance condition' as being involved in interpretation, since one could in principle interpret a formalism (or a score) in some way not intended: e.g. having a representation relation that maps musical scores to colors rather than sounds (or the *wrong* kinds of sounds: penny whistle rather than full orchestra), or a relation that maps a mathematical formalism to states of an abstract computer programme rather than a (real or possible) universe, or one of its subsystems. Once we have the interpretation, we can to a large extent ignore the physical world and focus on the interpreted theory itself, making discoveries in it that one would expect to find in the world – to test the theory will obviously require a comparison with observation and experiment, but it is likely true that most theoretical and mathematical physicists spend more time looking at a whiteboard or pad of paper (on which they construct and 'explore' some representation) than the very world that is their real target.

The multiplicity of interpretations is a blessing in the musical context: imagine how dull life would be if all performances sounded the same! But in the case of a mathematical formulation of a theory that admits multiple interpretations (different ways of filling in the semantics), we face a problem if we view our theories as telling us how the world is: these interpretations are supposed to be telling us how the world is, and yet they are usually incompatible. One interpretation might be local, so that there is no action-at-a-distance, while another interpretation (of the same theory) might be nonlocal. One might be deterministic, another indeterministic, and so on. They can't all be viewed as gripping onto our world since there are contradictions involved. So here we have a striking dis-analogy with the musical case: we might prefer one performance over another, but we aren't forced into believing in 'the one true performance' as we are with a theory that aims to provide a picture of reality. This possibility of several incompatible ontologies constitutes one of the thorniest epistemological problems faced by philosophers of science: the problem of underdetermination.

There is an additional level to the interpretation game that we have not yet mentioned. This is that the interpreted formalism can also be subject to further interpretation (and further multiplicity in interpretation). In other words, we view our object-to-be-interpreted not as a 'bare formalism,' but as an already-interpreted structure. For example, we might be studying the 'many worlds' interpretation of quantum mechanics (itself one of many observationally equivalent interpretations of the basic principles of quantum mechanics). We can ask of this interpreted formalism: what can the world be like for this to be true? Again, there might be multiplicity at this higher level, where supplying a semantics to the basic formalism is not enough to fix a world-picture (or all of its details). Are the many-worlds really to be viewed as separate universes that literally branch off from one another, or as something else not involving a literal branching of worlds? Likewise, with general relativity (Einstein's theory of gravitation) where we understand it to be a theory of spacetime geometry, the question arises again: given this interpreted formalism (in terms of spacetime curvature), what is the world like? Is there a literal spacetime geometry in the world, as substantial as tables and chairs? Or is it somehow built up from tables and chairs, and other forms of mass-energy. How much of the spacetime geometry picture do we suppose maps onto the world: just the geometrical structure or the extensionless points underlying this structure too?

Sometimes we simply call the initially interpreted formalism the theory, rather than an interpretation. For example, to most people (physicists included) general relativity is just a theory about the curvature of spacetime geometry and the way the curvature depends on the goings on of matter and energy in the spacetime. But this is only one possible approach. A flat spacetime picture with gravity mediated by an 'exchange particle' (the graviton) can also rightly be called general relativity. Here, the dynamically varying metric tensor (representing the geometry of spacetime in the orthodox picture) is treated as 'just another field in flat spacetime' – there are technical details involved in recovering the same predictions and symmetry properties, but we need not concern ourselves with these. The point is, we can treat the theory as involving curved space or flat space; that we tend to work with one approach does nothing to detract from the fact that the curved space picture *is also* an interpretation of sorts.

1.4 Why Does Physics Work?

One of the most puzzling facts about physics is that *it works*! How do theories perform this amazing feat. Easy, you might say: they work because they're *true*. But given the issue of multiplicity in interpretations of a theory (mentioned in the previous section), *which* picture is true? One often has very different ontologies (e.g. one with fields and one with particles instead; one with flat space, one with curved space instead, and so on). What concerns us here is why *mathematics* (an abstract thing, according to most) is able to predict and describe physical events. Why don't crystal balls, tea leaves, or any number of other things work? At least they are physical entities!

This problem was famously posed by Eugene Wigner in 1960 in a paper called "The Unreasonable Effectiveness of Mathematics in the Natural Sciences." He put the puzzle in strong terms: "the enormous usefulness of mathematics in the natural sciences is something bordering on the mysterious and that there is no rational explanation for it" ([56], 2). For example, there are branches of mathematics that allow us to predict the behaviors of planets, comets, missiles, and even certain elements of social systems, such as traffic flow patterns and queuing behavior. Why are the laws of physics so well-couched in the language of mathematics? Why do mathematical structures find such fruitful application in physics?

There are two kinds of application we can think of here, one more 'unreasonable' than the other: (1) a 'physics-dependent' kind, and (2) a 'physics-independent' kind. For example, there are cases where mathematics has been developed hand-in-hand with some piece of physics: the calculus was constructed with a physical problem in mind, namely how can we solve equations of motion (how do we know how a system will evolve in time). John von Neumann's creation of Hilbert space (in which quantum states are represented) was also of this kind: a case of *finding* appropriate mathematical tools for the job. The effectiveness of mathematics is clearly not so unreasonable in such cases: it is a criterion of successful tool-finding that it be an effective one. There were many tools not fit for purpose that were discarded.

The unreasonable kind is that application of mathematics in science (Wigner has in mind physics, primarily, but it generalizes to any mathematically modeled subject matter) that was created independently of physics and yet *later* found application in physics. Hence, the mathematicians were busying themselves with some purely mathematical problem, not caring a jot about the world of physics, and yet lo and behold this piece of mathematics is found to be a perfect fit for (some aspect of) the physical world. For example, complex numbers find a perfect home in quantum mechanics, as we will see, despite the fact that complex numbers were developed hundreds of years before quantum mechanics was conceived. Non-Euclidean geometries were found to provide the perfect framework for general relativity. The so-called 'spinors' of Henri Cartan, from 1913, were found to fit perfectly the intrinsic spin of electrons discovered in 1926 (and were pivotal in the theoretical prediction of anti-matter by Paul Dirac, who combined spinors with the mathematics of quantum mechanics). Wigner's own example involved a pair of old friends discussing one friend's job as a statistician. The other friend is incredulous to see that π (something to do with the ratio of circumference of the circle to its diameter) is appearing in a discussion of the population (humans!) via the Gaussian distribution. How can this be?

There are various responses one could give. One possible response is that even the 'purest' (most physics-independent) mathematics comes from worldly investigations at some level, as Stanisław Ulam points out:

> Even the most idealistic point of view of mathematics as a pure creation of the human mind must be reconciled with the fact that the choice of definitions and axioms of geometry – in fact of most mathematical concepts – is the result of impressions obtained through our senses from external stimuli and inherently from observations and experiments in the 'external world.' ([51], p. 284)

This doesn't really explain how mathematics can extend physics to go beyond what we can obtain through the senses: our senses are often wrong, and certainly can't put us directly in touch with atoms and quarks.

Better, one might deflate the puzzle by showing how it is no more miraculous than finding, e.g. a piece of furniture that fits 'just so' into some space in a room – and to think, the manufacturer had no idea about my room! I don't wish to suggest that mathematical physics is on a par with interior design, of course, but in terms of the basic principle behind the deployment of mathematics in physics, I think there are parallels here. But there is a crucial aspect left untreated: the furniture fits because the space has a structure with the right dimensions. It is a matching of (aspects of) their structures that grounds the fit. What about the case of mathematics and physics?

To answer this we might adopt the view that since mathematics is a science of patterns and structure and the world is patterned and structured, there is no mystery about their relationship: they simply have the same structure. Quite naturally, some structures will match up and others won't: the unreasonable effectiveness then just amounts to finding isomorphic structures, and why should it matter that the mathematical structure was discovered before the physical structure? Further, in physics, one is often dealing with a very limited list of physical features to be represented mathematically, so a matching between them is not so far fetched.

On this view, when one has a match between some piece of mathematics and some aspect of physical reality, then one has made a discovery that the world *has* this mathematical structure. One can then perhaps explain why the mathematics can lead to surprising physical discoveries – situations where we appear to get more out of the mathematics than we put in, to use Wigner's expression. A problem with this view is that sometimes mathematics is effective without us wanting to say that there is some structural isomorphism between the mathematics and the world. For example, we might have a model that would be *physically inconsistent* (such as the early model of the orbit of the electron around the nucleus, which predicted a rapid collapse of the atom) and so couldn't possibly be matched by reality. There are also all sorts of 'idealizations' in physics in which the mathematical model and the world can't be seen to correspond in terms of actual structure.

It is a good idea to keep in mind that we are dealing with scientific representations – most often (if not always) *mathematical* in the context of physics. The aim, as mentioned in the previous section, is to 'mimic' (or capture) in the representation the key (relevant, important) features of the system you are interested in (the target system). One can try to visualize what is going on in such modeling in fig. 1.3.

Here we start with a real-world system, and construct some model (which will neglect many of the details of our intended target in the world: transforming a real cow into a 'spherical cow' for simplicity, for example). We then deal with the model system and develop a theory of its behavior. The question is whether the theoretical representation describing the model also describes the worldly system – in terms of fig. 1.3, we ask whether the diagram *commutes*: must we go via the paired down model system, or can we use the representation directly for the worldly system? In terms of the success of mathematical representations, the miraculous quality concerns the fact that such representations often allow us to perform this 'bypass operation,' using the representation to describe and predict the behavior of various systems of interest, when the mathematics was developed in an entirely different context.[5]

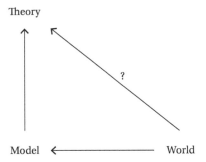

Fig. 1.3   A diagram making the problems of success (and failure) of the application of mathematics in the scientific enterprise a little more transparent. A real-world system is turned into a 'model' system (by abstracting away complex, unnecessary details), which is then involved in the theoretical description. If this phase is successful then we can view the theory as describing the real-world system. If not, then there may have been some problem in the model-building/representation stage (perhaps an over-idealization), so that 'Model' and 'World' are too dissimilar (in some relevant way) for a theory about the model to apply to the real-world target.

1.5 Further Readings

There are several philosophy of physics textbooks that I warmly recommend, though some are pitched many levels up from the present book. Those that are pitched at roughly the same level, adopt different approaches, tackle different kinds of question, or are less general.

Fun

- Nick Huggett (2010) *Everywhere and Everywhen: Adventures in Physics and Philosophy*. Oxford University Press.

  - This is a particularly sparkling treatment of several interesting themes in philosophy of physics, mostly focusing on space and time.

Serious

- Lawrence Sklar (2010) *Philosophy of Physics*. Oxford University Press.

  - Still a very good philosophy of physics introduction, though starting to show its age a little.

- James Cushing (1998) *Philosophical Concepts in Physics: The Historical Relation Between Philosophy and Scientific Theories*. Cambridge University Press.

- Wonderfully wide-ranging discussion of the intersection of philosophy and physics (stretching back to ancient physics and advancing all the way to general relativity and quantum theory). A great way to get some history, physics, and philosophy in one place.

Connoisseurs

- Jeremy Butterfield and John Earman, eds. (2007) *Handbook on the Philosophy of Physics*. Elsevier.

  - This represents a professional-level treatment of various subjects in philosophy of physics, written by a team of authors that includes physicists. Mastering the chapters of this volume should be something you aspire to!

- Robert Batterman, ed. (2013) *The Oxford Handbook of Philosophy of Physics*. Oxford University Press.

  - This collection of new essays provides introductory treatments of more modern debates in philosophy of physics, including especially condensed matter and statistical physics. However, it includes very good chapters on quantum mechanics, spacetime, and other more standard topics.

2 General Concepts of Physics

This chapter introduces, very briefly, several core concepts of physics from a bird's-eye view, rather than up close in the context of specific theories (something we do in later chapters, for classical and quantum, statistical and non-statistical, and relativistic and non-relativistic physics). Here we find out about *states* (a full specification of a system's properties, or values of all variables, at some instant), *observables* (those variables in a theory that can be measured and given a physical interpretation), and *dynamics* (the rules governing the behavior of a system, e.g. under the action of *forces*): the three core features that go into the construction of a physical theory and that form the raw materials for our interpretations. One finds these same basic concepts replicated across the theoretical frameworks above, where their specific realizations will differ according to the nature of the systems the theory is supposed to describe. These concepts are, then, at the root of the mathematical representations that we wish to make physical sense of. In particular, differences (of interpretation) can be seen to emerge within a theory about what kind of stuff the states and observables refer to. The dynamics enters this same interpretative debate in a variety of ways, especially in virtue of its link to symmetries – the next chapter puts these three basic concepts (*states, observables, dynamics*) to work in unpacking the concept of symmetry that will figure heavily in the remainder of the book.

2.1 The 'Three Pillars'

Philosophers of physics often like to speak of the 'pillars of modern physics,' by which they have in mind the theories of relativity, quantum mechanics, and statistical physics. What they mean is that these three together provide frameworks for the rest of physics – whether they are directly employed or not, they are seen to underlie all other phenomena (see fig. 2.1). However, one shouldn't take the 'pillars' metaphor too literally: they do not stand in isolation like architectural pillars, linked only by what they support or rest upon. Rather, as we will see, they overlap considerably. Perhaps a better metaphor is to think of them as distinct strands of fabric woven together to make up a single sweater (a very nice one, not like your granny might knit).

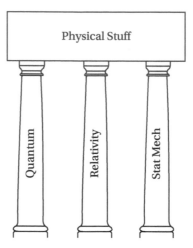

Fig. 2.1   The three pillars of modern physics: all phenomena of nature are viewed as reducible to three basic frameworks: statistical physics (often called 'stat mech'), the theories of relativity (special and general), and quantum theory – a common, though rather inaccurate picture of modern physics.

The three pillars are all examples of *spacetime theories*: they include spacetime (or space and time) as one of the fundamental elements of reality. Spacetime (or space and time) is part of most representations employed in physics: if they are supposed to model the world, then they should contain space and time because this is how the world seems to be configured, at least at some level of approximation – some theories of 'quantum gravity' suggest spacetime is *not* a fundamental feature of reality, so that spacetime 'emerges' from some deeper non-spatiotemporal theory: we briefly discuss quantum gravity in §§8.4 and 8.5.

Spacetime theories tend to match up with respect to their basic (deepest) structure: a set of points taken to represent the basic events of reality (or locations where events, such as colliding point particles, can take place) – however, some have argued that independently of further structure such 'bare' points can't represent real physical stuff. Distinct theories then diverge according to what further structure is applied to this foundation, depending on what they wish the theory to represent. Onto this set of points we can lay 'charts' or coordinates, labeling them and allowing us to speak of the points' relationships to one another – this set of points has the structure of a 'manifold,' namely something that 'looks locally' (i.e. at short distances) like ordinary flat space, but can vary in all sorts of ways globally (think of a newspaper laid flat versus rolled up: the way they differ is said to be a 'global' difference, but viewed up close enough there is no way to see such differences). We can map these points or regions of a manifold to itself (via transformations) to represent all sorts of possible changes (spatial movements, rotations, time evolutions, etc.) that might occur in the universe thus modeled (or in our observations of that universe) – or, more importantly (as we see in the next chapter), we can see what *stays the same* (is invariant) as certain such changes of the manifold are made. Such invariances are the stuff of laws of nature.

Mathematical structures, capable of living on this manifold (or a more complexly structured space, with a metric enabling talk of distances perhaps) are chosen with care to match features of the properties and behaviors of objects being described. We need to establish a matching (an isomorphism) between the way the chosen mathematical objects transform and the way we think the systems represented transform. For example, the 'physical things' (systems) of a theory (particles, fields, strings, etc.) are defined on this manifold structure and are represented by 'geometrical objects' (scalars, vectors, tensors, spinors, etc.). These correspond to the objects that we would think of as 'occupying' space and time (but this is really a matter of interpretation, as we will see). The objects are characterized by their behavior under mappings of the manifold, such as changes of the coordinates (corresponding to motion or rotation), as mentioned above. That such objects are defined relative to a spacetime manifold brings with it all sorts of nice mathematical tools and concepts from calculus and elsewhere, making the business of modern physics possible.

A point particle will occupy a single manifold (spacetime) point, fields infinitely many points (with a field-value located at each manifold point), and strings a one-dimensional manifold's worth of spacetime points (see fig. 2.2). This (manifold plus entity) gives us a preliminary set of elements for world-building: a set of objects locatable in space and time that might be relatable in various ways and that might have various possible

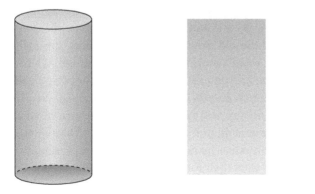

Fig. 2.2   A worldtube, worldsheet, and worldline, as generated by the time evolution of a disc, line (or open string), and point-particle respectively. Time goes up the page, and space across.

trajectories through the space. Note that we don't have to have our basic objects exactly equivalent to what we wish to model: there will always be approximations depending on the task. For example, it is perfectly possible to treat the Earth as a point in some model if all we need to think about is its position, say.

Still, much is missing in terms of representing a world like ours: we need to know more about what properties the basic objects have, how they combine and interact, and how (and why) they change and move. These require specifying the states and observables (roughly corresponding to *kinematics*) and their evolution over time (roughly corresponding to *dynamics*). This will supply us with a formal representation of a physical system (or possibly many systems, or even a whole universe or ensemble of such!). Referring back to the previous chapter, however, we find that this interpretative package (kinematics + dynamics) is rarely if ever *uniquely* determined by what we experience. For example, quantum theories can be supplied with radically different dynamics – e.g. ones in which measurement is central to the dynamics (causing the states to collapse to a definite value), and others in which measurement plays no such special role.

With such a representation to hand, we can ask all sorts of philosophical questions about the representation relation between model/theory and the world. For example, although it looks from the mathematical construction as though the spacetime points come first in ontological order, we should be careful in making such interpretive leaps. We can ask whether matter and spacetime are 'equally fundamental,' or if one is 'more fundamental' than the other. That is, if if we think our ultimate mathematical representation faithfully maps onto the world, we should be careful about confusing the order (or hierarchy) of construction of the representation with a corresponding order in reality, or in believing that every aspect of the mathematical structure has a corresponding target in the world.

There are other traps lying in wait that might be generated by the mathematical representation, yet without corresponding elements in reality. The modern version of the old pre-Socratic debate about the reality of space and time (and its relationship with matter) arises here. We can ask how symmetries of space and time act on physical situations and whether the new states they generate are physically real in this sense. We can ask whether spacetime points are real (despite the appearance of what are taken to be spacetime manifold points in the mathematical model). And so on. The point should be clear by now: mathematical representations of physical systems do not wear their interpretations on their sleeves – it will be even clearer by the end of this book . . .

In the next section we lay out the above-mentioned basic elements of a physical theory: <K>states, observables, dynamics<L>. This triple essentially packages together the 'kinematics' (states + observables: also relating to space, time, and motion) and the 'dynamics' (the physical forces and interactions constraining the kinematically possible motions), viewing a theory as the combination of these, thus making an interpreter's life easier by giving us the systems and their properties along with a rule (the dynamics) for how they change and vary over time and space.

2.2 Kinematics and Dynamics

Space, time, and motion (of some basic objects) are the central elements in the *kinematics* of a physical theory. Usually, this basic background must be decided upon first, and then the laws (dynamics) will be introduced to constrain what motions are *'actually* possible' relative to such a background. The division into kinematics and dynamics comes to us from Aristotle who viewed *kinesis* as a kind of 'potential' state of being while dynamics was an 'actual' state of being. Historians have wracked their brains over this distinction of Aristotle's for many centuries, but translated into our terms we can see that kinematics concerns possible motions when we ignore the action of any forces and laws of nature in the spatiotemporal background, while dynamics concerns what motions can be actual once the laws (such as Newton's laws of motion) are introduced. Mechanics is classically understood to be a fairly straightforward combination of these two components: kinematics + dynamics. All features of a world are understood to flow from a specification of both elements.

The kinematically possible trajectories will of course *include* the dynamically possible trajectories: the former space of possibilities is far larger than the latter. The modern distinction can be linked fairly closely to Aristotle's (from the previous chapter) by focusing on what is possible in the two scenarios: kinematics is about which motions are possible given the constraints of the spacetime itself along with the barest features of the basic objects (so that, for example, in a world with three dimensions, motions requiring more dimensions will not be kinematically possible). We can think of these as metaphysically possible worlds, but not necessarily physically possible worlds: worlds that are *conceivable*, but are perhaps not compatible with our laws. Physical reasonableness is the province of dynamics, which narrows down the space of metaphysically possible worlds to a family of physically possible worlds: worlds that are compatible with our laws. We can think of the introduction of dynamics as a demand for explanation concerning why things change their motions (or stop): this demands *forces* (and we have the law of inertia, embodying the tendency of bodies to stay in motion unless *forced* to do otherwise). Hence, we have the standard conception of kinematics as the study of the motions of bodies in the absence of forces, and dynamics as the study of the effects of forces on those motions.

Though we utilize mathematics in this representation of trajectories (motions), especially using geometrical notions, there is a radical disconnect in how the physically applied concepts relate to the pure (mathematical) concepts. For example, a *motion* in the geometrical sense simply involves *associating* one point to another point, with no sense of a continuous trajectory linking them (at least not of necessity: one could *imagine* the point being carried along in a continuous path, but it is not essential). In the case of physical motions, however, the smooth paths between initial and final points are crucial, and form part of our picture of how the world works – not least because we often need to know the duration of the time interval during which some continuous path was traversed. But deeper than this (though not undeniable) is the belief that in order to get from a point *A* to a point *B*, the points in between must be traversed, during which the object that moves retains its identity in some sense (and so is the same object at *B* as it was at *A* – a relation sometimes called 'genidentity').

In modern approaches to physics, we do in fact shift to a more abstract representation: we speak of 'states' and 'observables' in place of kinematics (with its associated space, time, and motion). But we don't fully dispense with space and motion; rather, a different kind of space and motion is employed, in the form of a *state space* and trajectories in this space. Just as we might build up a space from all possible combinations of some parameters, such as *x*, *y*, *z* measurements for ordinary space, we can also view the so-called 'canonical variables' as 'generalized coordinates' for this new kind of space (known as *phase space*: the state space of classical mechanics). Each point represents a different assignment of position and momentum to a system. This lets us do things we can't do in ordinary physical space. For example, when we are dealing with a complex system of many particles (with a large number of particles, *N*), it would be a complicated task to deal individually with each of their paths through three-dimensional space. But with state spaces we can bundle all of this information into a new space of $6N$-dimensions, in which a single point represents the positions (a 'configuration') and momenta of all *N* particles (taking into account each particle's three spatial coordinates in ordinary space and their momenta in three spatial directions) taken at an instant of time.

Let us fill in some of the missing details from the above account. The *state* of a system, as the name suggests, is a snapshot of a system, containing complete information about it at an instant of time – the system itself is understood to be well represented by this state, which is essentially built from its properties. In classical mechanics this state is simply the position *q* (in 'physical space') of a particle together with its momentum *p* (the 'canonical variables' from above). The idea is that from such a specification of a state, we can run it through the laws of the theory to get to the system's state at any other time (the dynamics in this new scheme). The state is, in this sense, the input (the programme) for the laws (the processor: a certain set of equations of motion known as Hamilton's equations – basically, Newton's laws of motion rewritten in fancier mathematics!), which separate the physically possible evolutions from the impossible ones.

Physics, at its most general, is concerned with physical quantities (in ordinary language we would call them 'properties'), the interactions between quantities of the same and different kinds, and the rate of change of such quantities. Once we have our set of quantities, which we'll label $\mathcal{A}$ (i.e. the observables: the things we can measure), that will define the *instantaneous* state of the object that interests us (and, in a sense, is how an object is defined in physics), we can think about how these quantities (and so the state, and so the object) might change *over time*. We can set up equations of motion of the general form:

$$\text{rate of change of } \mathcal{A} = \text{function of } \mathcal{A}. \tag{2.1}$$

Finding how a system will evolve then simply amounts to finding the particular function, which results from the investigations of physicists. Integrating both sides of the equation, we can find future (and past) values of $\mathcal{A}$ from its present value. (However, as we will see in Chapter 7, in quantum mechanics, as standardly interpreted, this framework only holds while the system is *unobserved* (between observations); measurement instead delivers a random value from a distribution of possible values, known as *eigenvalues*. In other words, there are *two kinds of dynamics*. Naturally, it would be better to make do with one, in terms of the number of elements in one's world-picture, and such interpretations exist, as we will see.)

The position and momentum above are *observables* of the system, and we find that specifying all of the values of a system's observables will uniquely determine its state – likewise, knowing the state means knowing the values of the observables. We measure observables to gain knowledge about the state. Hence, we have a perfect correlation between states and (complete sets) of observables at an instant t—$(q(t), p(t))$ are complete in classical mechanics in the sense that all other observables can be constructed from them using some mathematical operations. Such observables provide the core link between theory and world in the context of physics: they are the things we measure and whose values we predict. As such, they ground the qualitative character of a world: two worlds that are exact duplicates in terms of all of their observables will thereby be qualitatively indistinguishable (a feature that will be important in later examples). (The observables of classical mechanics have a *dual* role: on the one hand they are measurable quantities, allowing us to latch the theory onto the world, providing information about a state of the world. On the other hand they generate specific transitions of the state of a system from one to another: for example, the energy observable (aka the Hamiltonian function), when viewed in phase space terms, generates time-translations.)

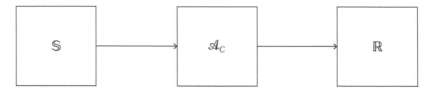

Fig. 2.3   Representation of an observable $\mathcal{A}_C$ in classical physics, mediating between an abstract state space $\mathbb{S}$ and some numerical value that is associated with an experimental outcome $n \in \mathbb{R}$.

The observables in the classical case are functions from the phase space to (real) numbers: this is the link between the abstract representation, given by the state space, and reality, as given in measurements that can be associated with real number values. Take as an extremely simple example (a standard case study for physics introductions): a coin. This is a two-state system, and so it has a state space with just two points: *Heads* and *Tails*. An observable $\mathcal{A}$ would have to be a function on this space, so that it spits out some numerical value depending on what state it is fed. We can write $\mathcal{A}$ (*Heads*) = 1 and $\mathcal{A}$ (*Tails*) = 0 – now we have made this association, if we want to be formal we can write the state space as $\mathbb{S} = \{0, 1\}$. Far more complicated examples occur in physics, such as energy observables that spit out a number representing the total energy of a system (that can deliver a continuum of possible values, rather than $\{0, 1\}$) when fed a point from the phase space. But still, this simple setup gets the basic point: classical observables are functions from the state space to numbers: $\mathcal{A} : \mathbb{S} \to \mathbb{R}$ (see fig. 2.3).

Quite naturally, the system's state space will depend on the kind of system it is. Quantum mechanical systems are so radically different (e.g. not generally possessing sharp values for their observables, but instead a 'spread' of possible values called *eigenvalues*, which have probabilities assigned to them, or to their being found on measurement), however, that a different setup is required. The position and momentum observables that have real number values in the classical context are understood to be *operators* ('Hermitian matrices' in the jargon) on states in the quantum context. However, the eigenvalues of the operator (representing measurement outcomes) *do* have real number values, and these are what we measure in experiments (see fig. 2.4). These operators, as with the classical observables, also serve to create new states from old by their action.

As we will see later, much of this difference has to do with the fact that quantum particles have both wave and particle aspects. For this reason, the state of a system is represented by a function $\psi(x, t)$ (a 'wavefunction') that represents the *amplitude* of the wave aspect of the system at the location x (given by three spatial coordinates) at time t. This wavefunction must be a complex number in order to represent the interference effects

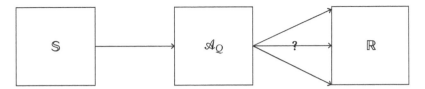

Fig. 2.4   Representation of an observable $\mathscr{A}_Q$ in quantum physics, mediating between an abstract state space $\mathbb{S}$ and a range of numerical values (eigenvalues), each associated with a *possible* experimental outcome $n_i \in \mathbb{R}$

usually associated with wave phenomena and so is not itself an observable quantity. But one can construct a real-valued quantity that *is* observable by taking its squared modulus (the wave's 'intensity'): $|\psi|^2$ – this is the 'complex square' involving multiplication of the complex number by its complex conjugate: $|\psi|^2 = \psi^*(x, t)\psi(x, t)$. If you want to know the probability of finding a particle at some particular location x at time t, $Prob(x, t)$, it is this squared modulus that you need to invoke. If we make the identification $Prob(x, t) = |\psi|^2$, then if all the possible outcomes for the location are integrated together (i.e. integrate over an interval, say from a to b), then, since probabilities must always sum to 1, we have:

$$\int_a^b \psi^*(x,t)\psi(x,t)\,dx = 1 \quad \forall t. \tag{2.2}$$

For the dynamics, given in schematic form in eq. 2.1, in the case of quantum mechanics, we need an equation for the behavior of this wave-function ψ *that is written as a function of* ψ: this is the Schrödinger equation. The state space in which the various ψ are represented is known as 'Hilbert space' (a kind of vector space) in which states are represented by 'rays' (basically, vectors with some redundancy removed) in the space. This has the appropriate properties to represent the observed wavelike features of quantum systems. The observables (operators) are then understood to be objects that act on this space (simply: matrices acting on vectors) to produce numbers (the eigenvalues) comparable with experiment – these matrices must be Hermitian (self-adjoint, or equal to their own conjugates) precisely to get experimentally measurable real numbers out.

Hence, though there are very significant differences in the specific mathematical objects used to represent the states, observables, and dynamics, the same basic structures (harking back to the distinction between kinematics and dynamics) for representing the world and our engagement with it are utilized – ditto in statistical physics (which can be classical or quantum). However, we will find in later parts of this book that some pressure is placed on the ordering of the triple, ⟨(states, observables), dynamics⟩, since in general relativity the dynamics need to be solved first before sense can be made of the kinematics: spacetime geometry (needed for the definition of kinematics) comes out of solutions to the equations of motion (the dynamics) of general relativity – the slogan is 'no kinematics without dynamics'.

2.3 Reference Frames, Invariance, and Covariance

In a great many cases, we can determine the basic features of a theory by invoking certain 'principles' on which the theory is based. Principles of physical theory are supposed to be claims about the world that are somehow *more robust* than most other such claims about the natural world. They say things about the world that are very hard to imagine not being true. In other words, they are as close to universal (empirical) truths as one can get in physical theories. Such principles are really 'meta-laws' (laws about laws). Special relativity, for example, involves two core principles:

SR1   All inertial frames are equal (from the point of view of mechanical and electromagnetic physical quantities and laws).
SR2   The speed of light is constant (in any and all inertial frames).

Or we have the single principle of Galilean relativity satisfied by non-relativistic laws:

G   The laws of motion have the same form in all inertial frames (from the point of view of mechanical physical quantities and laws).

In the case of thermodynamics, one has:

T   The laws should not allow the creation of perpetual motion machines.

These laws govern laws: if a theory is to describe specially relativistic systems then it must contain only laws that satisfy the two principles, SR1 and SR2, above. In a sense, the principles constitute *what it is to be* a specially relativistic (or Galilean relativistic or thermodynamic, and so on) system. While the 'normal' laws of a theory might involve a reference to some specific type of system (particular particles or fields, for example), these meta-laws float above such details: they are far more general, and therefore also more robust (i.e. to changes in the specific details of the theories that implement the principles). For example, special relativity was devised before quantum mechanics came about, yet it applies just as well in quantum mechanics as it does in classical physics.

Often, as seen above, these principles and laws concern the extent to which the reference frame, from which observations are made, is arbitrary. A reference frame can simply be understood to be a set of coordinates (x, y, z), which we can think of as a spatial frame in which measurements will be 'recorded,' and a time coordinate t, which we can think of as the reading of 'clock time' for these measurements. This is the laboratory of physics, though here it is presented rather abstractly.

Laboratories will usually vary between observers. For example, experiments performed on the international space station will occupy a different frame of reference to yours in your office or wherever you are reading this book. One can imagine other experiments taking place in a laboratory that is rotating (i.e. spinning on its own axis), which would involve a different reference frame. In order to make sense of these differences we need some way of relating them to one another, so that arbitrary features of the reference frame are not mistaken for features of reality. Transformation laws link the various reference frames, and allow us to see that the physical quantities that we measure do not depend on the frame in which they are measured: we *extract* the physical structure as that which is left invariant between the frames. As we see in the next chapter, these amount to *symmetry principles*.

These high-level laws are then defined by invariance with respect to some class of transformations (some way of shifting, rotating, evolving, twisting, or otherwise morphing the system), in which case we say that the laws in question are *covariant* with respect to those transformations – with the systems themselves then said to be *invariant* rather than covariant (the relevant quantities of the systems, such as energy or momentum, are then said to be *conserved*). In the case of special relativity above, the task was to find a set of transformation laws that *preserve* the principles (given that those principles seem to have a solid status in reality, as revealed by experiment). These are known as the Lorentz transformations, which Einstein encoded into the structure of spacetime (thus providing an entirely new kinematic framework for considering the motion of bodies in space and time). In the case of Newton's equations of motion one has to find transformation laws that preserve those equations – these are the so-called Galilean transformations (discussed in the next chapter: for now, think of these, and the Lorentz transformations, simply as ways of moving a system around in space and time). We can, given this, rewrite the principles from above as:

G　Galilean relativity → The covariance of the equations of motion (laws) under Galilean transformations.

SR　Special relativity → The covariance of the equations of motion (laws) under Lorentz transformations.

These relativity principles bring into center stage the specific reference frames as characterized by their invariance under some specific set of transformations. It is the invariance that is really key. We turn to these invariances (symmetries) in the next chapter.

2.4 Further Readings

There are several books that I wish I'd known about when I really started becoming keen on physics. I hope these will prove useful for those wishing to build a solid background in mathematics and physics.

Fun

- Robert Mills (1994) *Space, Time and Quanta*. W. H. Freeman.

  - Brilliant exposition of the basics of contemporary physics. For beginners, but manages to introduce many important mathematical concepts.

- John Taylor (2001) *Hidden Unity in Nature's Laws*. Cambridge University Press.

  - As above, elementary but written with a master's touch.

Serious

- Leonard Susskind (2014) *The Theoretical Minimum: What You Need to Know to Start Doing Physics*. Basic Books.

  - The perfect book for readers wishing to gain some facility in doing computations in physics.

- Lawrence Sklar (2013) *Philosophy and the Foundations of Dynamics*. Cambridge University Press.

  - Textbook considering in depth philosophical aspects of many of the issues considered in this chapter – historical details are nicely interwoven with the philosophy.

Connoisseurs

- Roger Penrose (2007) *The Road to Reality: A Complete Guide to the Laws of the Universe*. Vintage.

  - Do-it-yourself guide to becoming a theoretical physicist. It's a fairly bumpy road, but full of philosophical insights and superb, readable introductions to even extremely difficult concepts.

3 Symmetries in Physics

She said, "You're wearing two different coloured socks." I said, "Yes, but to me they're the same because I go by thickness."

Steven Wright

Modern physics is simply inconceivable without symmetry. As the previous chapter hinted at, symmetries are deeply entangled with the physical laws and conservation principles forming the roots of our best theories. In many ways these theories (and the entities and structures they describe) are *defined* by their symmetries. As we will see in the chapters that follow, symmetry also lies at the roots of many of the philosophical problems faced by physics: the interpretation game is made much more difficult (or more interesting) by the presence of symmetries.

"Symmetry" in days gone by (and present days by your average 'person on the street') had more of an aesthetic meaning: to do with being harmonious, balanced, or *well-proportioned* – this is the original meaning of the Greek word συμμετρια ('same measure'). This is a rather static sense: a figure, face, or building can be symmetric without our doing anything to it. But in more modern terms, a symmetry involves observing something before and after some *action* has been performed on it. It is a kind of 'change without change': *transformations* (or operations) that leave something (or everything) about an object the same as before. They point to an 'insensitivity' or 'blindness' of some (in cases that interest us, physically) relevant properties or relations (including laws of nature) to some transformation/s that might rotate, shuffle, move, twist, push, or otherwise modify a system in some way. In such a case, the transformations that this holds true for are symmetries, and the object (possibly a physical theory) is said to be invariant (or symmetric) under the transformation.

To pick up Steven Wright's joke about the socks above: clearly switching a blue for a red sock (a possible operation) will not be a symmetry in terms of how they look, but if we don't care about color, by focussing on thickness, then, assuming they *are* the same thickness, switching the socks will be a symmetry *in terms of thickness*: thickness is *preserved* or is left *invariant* with respect to the operation of switching them. Going by thickness determines a different equivalence class of 'the same things' (defined by their 'switchability' without altering some specific property of interest) than if we go by color.

This brings out an important feature of symmetries: they involve 'ignoring' some aspects of a situation and focusing on some *relevant* structure that is preserved during some operation. That is, when we speak of a transformed object's being indistinguishable from the original, we usually mean that it is indistinguishable in some relevant respects. In many cases that interest us in this book, it is the laws of nature (and the associated states and observables) that are left invariant with respect to some operation, even though some other aspects might be changed. Sometimes, however, we find that the indistinguishability concerns all relevant respects that would seem to make a physical difference. Such scenarios underlie many philosophical debates since there are both arguments for viewing the transformed states as physically distinct (despite their indistinguishability) and for viewing the transformed states as physically one and the same possibility (that is simply being represented in different ways).

3.1 Symmetry, Invariance, and Equivalence

Symmetries involve equivalence of some sort or another; this is their defining characteristic: two distinct things are equivalent with respect to some feature or features. Given that these things are usually seen to be physically distinct, the equivalence means that for the purposes of physics, 'either will do' for formulating some problem: they have the same information content. This translates into a claim about distinguishability, as suggested above: given an equivalence, one will not be able to distinguish (internally: without reference to some other features) which of a pair of symmetric scenarios one is faced with. For example, when sitting on a train at rest on a platform with another train sitting alongside also at rest, when there is some relative motion it is often difficult to tell whether it is your train or the neighboring train that is in motion. Externally, by looking at your position relative to the reference frame provided by the fixed buildings around the train, you can then distinguish rest from motion.

The following (rather lengthy) passage from Galileo's *Dialogues on the Two Chief World Systems* – certainly among the most famous pieces of writing in the history of physics – rests on a similar phenomenon:

> Shut yourself up with some friend in the main cabin below decks on some large ship, and have with you there some flies, butterflies, and other small flying animals. Have a large bowl of water with some fish in it; hang up a bottle that empties drop by drop into a wide vessel beneath it. With the ship standing still, observe carefully how the little animals fly with equal speed to all sides of the cabin. The fish swim indifferently in all directions; the drops fall into the vessel beneath; and, in throwing something to your

friend, you need throw it no more strongly in one direction than another, the distances being equal; jumping with your feet together, you pass equal spaces in every direction.

When you have observed all these things carefully (though doubtless when the ship is standing still everything must happen in this way), have the ship proceed with any speed you like, so long as the motion is uniform and not fluctuating this way and that. You will discover not the least change in all the effects named, nor could you tell from any of them whether the ship was moving or standing still. In jumping, you will pass on the floor the same spaces as before, nor will you make larger jumps toward the stern than toward the prow even though the ship is moving quite rapidly, despite the fact that during the time that you are in the air the floor under you will be going in a direction opposite to your jump. In throwing something to your companion, you will need no more force to get it to him whether he is in the direction of the bow or the stern, with yourself situated opposite. The droplets will fall as before into the vessel beneath without dropping toward the stern, although while the drops are in the air the ship runs many spans. The fish in their water will swim toward the front of their bowl with no more effort than toward the back, and will go with equal ease to bait placed anywhere around the edges of the bowl. Finally the butterflies and flies will continue their flights indifferently toward every side, nor will it ever happen that they are concentrated toward the stern, as if tired out from keeping up with the course of the ship, from which they will have been separated during long intervals by keeping themselves in the air. And if smoke is made by burning some incense, it will be seen going up in the form of a little cloud, remaining still and moving no more toward one side than the other. The cause of all these correspondences of effects is the fact that the ship's motion is common to all the things contained in it, and to the air also. That is why I said you should be below decks; for if this took place above in the open air, which would not follow the course of the ship, more or less noticeable differences would be seen in some of the effects noted.

In refreshingly simple language, this passage destroys the intuitive reasoning for thinking that we live on a stationary Earth: why don't we feel the wind in our faces, and why aren't we thrown off the surface?! Surely, the critics said, we should *see* and *feel* that it moves? By linking the Earth whizzing through space to a ship cutting unhindered through the water, Galileo is able to show how we would notice none of the effects we might naively expect to see and feel: the two situations would be internally indistinguishable, and so we could not detect which were true without using some kind of external check. Indeed, for the most part, the objects *external* to the earth with respect to which we might notice the existence of motion, are so far away that our motion is undetectable – if they were much closer (and/or our speed much (much!) faster) we would indeed notice the motion, and the ship–Earth analogy would break down (though with the experimental outcomes Galileo suggests none the wiser for it).

However, the passage also contains a symmetry principle (Galilean relativity) that expresses the idea that uniform motion is undetectable: the laws of motion are insensitive to transformations that involve switching a pair of uniformly moving reference frames. As Galileo puts it, there would be "not the least change" in the behaviors of objects in the ship. That is, the laws of mechanics (and any experiments you might conduct using them) are invariant with respect to changes of reference frame (in this case the ship) that differ by their (uniform) velocities and positions (and times). This is much easier to experience on an aeroplane (once it has reached a cruising altitude) or a train journey. So long as the going is smooth (and you don't sneak a peek out of the window), there is no way of telling that you are moving by observing the motion of objects in your vicinity. The air crew merrily stroll down the aisles with drinks, with no spillages – unless turbulence hits (which tends to occur precisely when the drinks are being served, but never mind that . . .), in which case the motion is non-uniform (this plays a significant role in issues of spacetime ontology, as we see in the next chapter). Hence, we have a change without change: an equivalence (of motions of objects below deck) coupled with a known physical difference (the different states of motion of the ship).

Galilean relativity amounts to the claim that the laws of motion are independent of location, time, orientation, or state of uniform motion at *constant* velocity (as in the ship example). Changing such features does not change the laws of nature (the same experimental results will emerge regardless of the state of uniform motion), and so these are symmetries of nature. The laws cannot be used to detect absolute locations, times, orientations, or states of uniform motion or rest. Any systems described by such laws are invariant with respect to those transformations that generate changes of location, time, orientation, or state of uniform motion (at constant velocity). This family of transformation-types gives us ten parameters describing ways a system can be altered while leaving the laws undisturbed (Nature's symmetry group in a classical mechanical world): three parameters to describe spatial translations, a single parameter for temporal translation, reorientations about the three axes of space (three more parameters), and velocity 'boosts' in each of the three directions of space (another three parameters). These are known as the *Galilean transformations*, and they satisfy the mathematical properties of being a *group* (basically a set of elements – in this case the various transformations – that must satisfy certain conditions on the way the elements combine).[1]

In more technical terms, for some initial state, represented by coordinates $(\mathbf{x}, t)$, we can consider transformations of coordinates of a system of the form:

$$(\mathbf{x}, t) \rightarrow (\mathbf{Ax} + \mathbf{v}t + \mathbf{b}, t + s) \tag{3.1}$$

The meaning of this is that we can act on some point (object) by: rotating it (represented by the matrix $\mathbf{A}$, which is formally an element of the rotation group $SO(3)$); by shifting it forward (evolving it) in time (represented by $t + s$, which, mathematically, is a simple shift along the real number line $\mathbb{R}$); by translating the point by **b** (where **b** is a vector in space $\mathbb{R}^3$); or by changing the velocity by some 'boost' **v**t (where **v** is a velocity vector in space). This gives us our ten-parameter group, the Galilean group, Gal for short. A theory that is invariant under these transformations is said to be Galilean invariant. What this means is simply that such transformations (singly, or in combination) 'make no difference' from the point of view of physical laws: if **x**(t) is some solution of the equations of a theory (that is, some trajectory of a system such as a particle) then so is the alteration $\phi(\mathbf{x}(t))$, with $\phi \in$ Gal. You can't use such transformations to 'test' what your absolute (i.e. relative to space and time, rather than other external objects) state of motion is. Note that $\phi(\mathbf{x}(t))$ and **x**(t) represent physically distinct solutions: they are different (as physical states: different positions, velocities, etc.) and yet they are also the same in terms of their intrinsic properties. In terms of the possible-world-talk from earlier, we can say that Galilean transformations leave a theory's space of physically possible worlds the same so that applying one of the transformations to a state (i.e. of a physically possible world) leads to another physically possible world (see §3.2 for more on the connection between laws and possible worlds).

Newton's laws of motion are independent of location, time, orientation, or state of uniform motion at constant velocity. Hence, they are Galilean invariant: they take the same form in reference frames related by the corresponding transformations. This implies that we can't detect position and velocity by any process governed by Newton's laws: a kind of 'conspiracy of nature' hides any such transformations from our gaze. As we shall see, this conspiracy lies at the center of a justly famous debate between Newton and Gottfried Leibniz. In situations where the solution concerns the entire material content of the universe, performing a Galilean transformation appears to generate something that *doesn't* differ physically: yet if we think of space and time as giant containers in which the transformations are carried out then they must be physically distinct regardless.

At the root of the debate is the fact that for each set of such parameters there is some piece of physical structure that is rendered *unobservable*. In the case of the translation piece of the Galilean group (the bit that shifts objects in space), one cannot observe absolute locations (i.e. positions relative to space itself). Invariance under time shifts renders absolute time location unobservable (you can't tell *when* you are relative to time itself). Invariance under rotations renders absolute orientation unobservable (you can't tell which way you are pointing relative to space itself). And, finally, invariance under changes in velocity renders absolute velocity (or absolute rest) unobservable (you can't tell how fast you are moving relative to space itself). That's a lot of unobservable structure. In general, the more symmetries one has the more unobservable structure one introduces since such symmetries destroy our ability to discern differences between certain physical situations, and so we can't tell which of some set of possibilities we are in fact observing (assuming there is indeed a fact of the matter).

There is a standard method of removing this unobservable structure (formally at least), if we decide that keeping it is not to our taste, by 'quotienting' (or 'modding') out the symmetry. What this amounts to, in simple terms, is identifying all of the symmetric possibilities (those directly related by some symmetry operation), forming (and subsequently working with) an 'equivalence class' – an equivalence class of possibilities can simply be understood to contain elements any of which would be 'up to the task' of representing some physical situation so that there are many possible mathematical descriptions of the same situation. The symmetry transformations can then be seen to take us from one possible description to another possible description. The quotienting procedure then reduces this multiplicity of indistinguishable possibilities to a single possibility: the space of physically possible worlds is thus reduced in size. In a sense, as we see in the next chapter, the battle between Leibniz and Newton centered on just this feature: how big is the space of possibilities? Should we include indistinguishable possibilities or eliminate them?

Finally, just for completeness, we should point out that (in certain special important cases involving *continuous symmetries*) each of the sets of parameters is associated with a *conserved quantity* as well as some unobservable structure. Translation symmetry leads to conservation of momentum; time translation symmetry to conservation of energy; and rotation symmetry to conservation of angular momentum. If there weren't these stable features under the various operations, then one could use the alterations to detect location, time, velocity, and orientation. That we cannot do this also points to features of the spatiotemporal system itself: it must be isotropic (with no preferred direction) and homogeneous (look the same everywhere and everywhen). This is a general feature of symmetries: what kinds of transformation leave the laws unchanged are indicative of the underlying physical structures in or on which the operations are carried out. Naturally, the kinds of objects that are left most unperturbed under various operations (those that are most insensitive to a barrage of transformations that attempt to change them) are going to be homogeneous among themselves – we find this even with quantum particles, which are identical in terms of their intrinsic properties, so that switching them leaves the laws of quantum mechanics none the wiser. Likewise, what is observable according to the theory is equally unimpressed by such switchings – in which case the permutation of particles constitutes what is called an *automorphism* of the set of observables, mapping it back to itself (if the equivalence mapping is to another object distinct from the first then it is generically known as an 'isomorphism'). We discuss this curious 'permutation symmetry' of quantum objects in §7.5.

3.2 Symmetries, Laws, and Worlds

Jennan Ismael and Bas van Fraassen [27] claim that a physical theory can usefully be split into two main ingredients: 'theoretical ontology' and 'laws of nature.' The ontology is to be thought of in terms of a metaphysical possibility space (a space embodying what kinds of entities, properties, and relations are allowed by the theory). These possible worlds (being 'entire world histories') correspond to state space trajectories. Neither of these spaces is yet physical, in the sense of satisfying the dynamical equations of some theory (the laws, that is). The role of the laws is to then restrict this rather liberal metaphysical possibility space (or the associated mathematical state space) to a smaller family of *physically* (or, as philosophers of science say, *nomologically*) possible worlds (those in which the entities, properties, and relations satisfy the laws). Hence, only some of the possible worlds (a subset of the *space* of possible worlds) is associated with the physically possible trajectories.

This notion of (physical) possibility space is deeply entangled with both the definition of a theory, then, but also with interpretation: the provision of the ontology (the worlds) is equivalent to giving an interpretation (a way the world could be according to the theory). The idea is that a theory presents us with a mathematical space, with a certain geometrical structure appropriate to the system it is invoked to model, where the points of this space represent physically possible states of affairs for a system modeled by the theory. Symmetries can easily be realized in terms of these such spaces. If a theory admits symmetries then: (1) distinct points of the space will be related by symmetries; (2) these points will form an equivalence class (an *orbit* in the jargon); and (3) elements of this equivalence class will represent qualitatively identical possibilities: the symmetries simply map a set of individuals (in a domain) onto other individuals in such a way as to preserve the relevant relations and properties (this is the meaning of the term 'automorphism' from earlier).

This framework lets us now distinguish between two types of symmetry, which is important from the point of view of possible worlds (i.e. distinct physical possibilities): symmetry proper and gauge symmetry. According to the former, the distinct points related by the symmetry transformation are taken to represent physically distinct possibilities – that is, the representation relation is one-to-one between elements of the space (trajectories) and physically possible worlds. But in the case of gauge symmetries this direct link breaks down so that many trajectories correspond to one and the same physically possible world – the representation relation is therefore many-to-one (we discuss this curious symmetry further in §8.3). The idea is that we would have some equations of motion defining some theory, which will admit some symmetric solutions. This has important interpretative consequences since the latter gauge symmetries open up the possibility of treating the equivalence class itself as the object that we should be committed to (as mentioned earlier). This is how symmetries begin to complicate the interpretation game.

3.3 Some Important Distinctions

The study of symmetries in physics (and elsewhere) is buried in jargon. This is no bad thing: in this case it means that they are well understood, and so involve lots of fine distinctions in classifying them. In this section we briefly introduce some of the most important distinctions that will be invoked in the remainder of the book.

Geometric and Dynamical

A very special class of symmetries relates to the properties of space, time, and spacetime. These are known as 'geometric (or universal) symmetries' and they are the basis of the laws of many theories. Indeed, one can work backwards (from certain laws to certain features of the space and time) or forwards (from features of space and time to the laws). What is special about geometric laws is that since every system exists within spacetime every system is thereby subject to symmetries pertaining to spacetime: hence, they are universal.

There are other symmetries that apply to some particularities of the system that is invariant under them. For example, there is a symmetry according to which the laws of quantum mechanics are invariant under changes of absolute phase of particles (where the phase comes from the strange wavelike aspects of particles in quantum mechanics). Quite naturally, this is grounded in the nature of the particles and in the existence of their phases, which is a feature not universally shared by all things. These are known as *dynamical symmetries* since they refer to specific physical theories (or rather the forces or interactions these theories describe – usually post-classical ones).

Eugene Wigner (who named this distinction: see his book mentioned in the Further Readings at the end of this chapter) grounds it in a distinction based on *events*: the geometric symmetries don't change the events (by which he clearly means observable, qualitative stuff: 'phenomena'), but just their spatiotemporal locations, orientations, and motions. Moreover, the formulation of the invariances makes reference only to events and their correlations, independently of the laws or the specific constitution of the physical systems: all that matters is the independence of the events from locations (regardless of what the events are made up of) and so on. The dynamical symmetries on the other hand involve the laws of physics directly in their formulation, explicitly referring to the specifics of the objects and interactions.

This distinction can usually be mapped onto a further distinction between 'internal' and 'external' symmetries, where the former refers to the fact that the transformations are not spatiotemporal but instead refer to some 'abstract' internal space. Many properties of quantum particles show this independence from spacetime. Such symmetries are further from experience, of course, because experience demands at least some semblance of a spatiotemporal framework to *be* experienced. As such internal symmetries are inferred from data about the forces, where the process of inference is more elaborate than for geometric symmetries.

Continuous and Discrete

The spacetime symmetries are said to be *continuous* symmetries since each transformation can be built (or reached) by repeatedly applying an infinitesimal transformation. Hence, a rotation from some original position to a 40-degree turn does not involve a jump, but must involve an infinite sequence of intermediate tiny rotations. Such symmetries, involving the addition of infinitesimal in-between stages, are sometimes known as 'proper' symmetries, since they involve physically realistic transformations that we could imagine doing with real objects in real space. By contrast we have *discrete* symmetries that do not involve intermediate steps through the points of space connecting one state (the original) with another (the end point of the transformation). An example is reflection about some axis, which no continuous sequence of rigid (infinitesimal) motions can bring about. A reflection can turn a left hand into a right hand (look at your hand in a mirror!). This means that reflection symmetry involves a left–right symmetry – a mirror doesn't actually transform your right hand into a left hand of course, it just shows what it would look like if you somehow managed it. Time reflection is another example: what it means is that if a process in one temporal direction is 'allowed' (i.e. a solution of some equations of motion), then so is its time-reverse (where the 'movie' of the process is played backwards).

These are, unsurprisingly, known as 'improper' since it is hard to envisage such jumps occurring in physical spacetime. In other words, we can't transform a left hand to a right hand by performing smooth, rigid, continuous motions. (Clearly the 'rigidity' is required here since otherwise we could just 'squish' a left hand like plasticine to such a degree that it looks like a right hand.)

As with the symmetries appearing in the Galilean group, such symmetries are nonetheless expected to be obeyed by the laws and quantities of physics given assumptions about the isotropy and homogeneity of space and time. However, we will see that philosophical novelties appear in the case of discrete symmetry groups.

Local and Global

As the name suggests, a *global* symmetry is one that involves transformations that act in the same way at every point of space, so that in, e.g. the case of a translation performing it globally would mean shifting everything in the same way. This might be taken to mean that they only ever apply to *everything* (i.e. the universe), but that's a mistake: Galileo's ship argument involves a global symmetry, namely a uniform velocity boost. Problems emerge, as we see in the next chapter, when we apply such global symmetry transformations to the universe as a whole. It appears that while Galileo's transformed ships will clearly be physically distinct scenarios (e.g. only one's moving away from the shore while the other is anchored), in the case of shifted universes as a whole, it's hard to see that there is any difference at all: there is no comparable sea shore from which to judge the differences.

A local symmetry, by contrast, involves transformations that can be applied at spacetime points *independently* of the other points. If the spacetime has infinitely many points then this implies that the symmetry group is infinite-dimensional (has infinitely many parameters, compared to the ten of the Galilean group), since it can act differently at each one of those points. These local symmetries are associated with internal symmetries since they cannot be couched in terms of conserved quantities observable through the various correlations of events (as with the time-*interval* between events being observable rather than the absolute temporal location, pointing to time-translation invariance). General relativity has such a local spacetime symmetry based on $\mathrm{Diff}(\mathcal{M})$, the group of diffeomorphisms of a manifold (spacetime, that is) – these diffeomorphisms essentially just turn one manifold into another by moving the points of the spacetime manifold around. What this means, very roughly, is that the laws of general relativity (Einstein's field equations for the gravitational field) are invariant with respect to any smooth transformation of the points of spacetime that keeps the structure of the spacetime manifold intact (which means, again very roughly, that the transformation does not end up 'tearing' the spacetime manifold or putting holes in it). To connect up to the discussion about laws, symmetries, and possible worlds: if we have some solution of the Einstein equations (which represents a possible world), then if we transform it by applying one of these diffeomorphisms we have ourselves another possible world that satisfies the laws of general relativity. In this sense, the group of diffeomorphisms is the symmetry group of the theory.[2]

Local symmetry is rather more surprising: clearly if we shift everything in the same way by the same amount (moving everything we can see, including ourselves, by the same amount) then we won't be surprised to see that things look the same. But if we shifted *everything* in *different* ways at all the different points (so that the transformation is dependent on the particular spacetime point), then we'd surely be very surprised if things (observables and laws) stayed the same! Local symmetries have just this property. Since they lead to no observable changes, they are usually considered to be a feature of the mathematical representation used to model reality, rather than part of reality itself, as with the symmetries used in Galileo's ship – though there is controversy over this point, some of which we return to. Such symmetries are also called *gauge* symmetries – a topic we return to in §8.3.

Passive and Active

Transformations (otherwise known as 'mappings' or just 'maps') in the sense used in this book simply refer to an association of one object (e.g. a point in spacetime) to another (e.g. another point in spacetime) – the new point that results from the transformation is the *image* under the mapping. Importantly, whatever is sitting at such a point is 'carried along' as if on a wave to the image. Because of this joint effect, we will generally be loose when talking about what gets mapped around. Symmetries are quite clearly mappings in this sense and can transform solutions of a theory's equations into other solutions (e.g. carrying a ship in one region of space to another region of space).

We can consider these transformations in either a 'passive' or an 'active' sense. The former means that it is the reference frame of the observer that is transformed – for example, I might stand on my head to see if some property (or all properties) of an object are left 'looking the same.' The latter means that I save my poor head and turn the object on its head instead, still checking to see if the properties are left unchanged or not. When we are dealing with a symmetry then the active and passive interpretations are equivalent: what the active transformation does is exactly compensated for by an equal and opposite (i.e. inversely related) passive transformation: doing an active followed by a passive transformation cancels out so that nothing ends up being done.

The simplest way to see that this is so involves looking at a transformation through a coordinate system. Let's consider simply rotating a ball, with starting coordinates (F_x, F_y), by θ (some number of degrees of rotation) relative to an origin. On the active interpretation, this would involve a literal

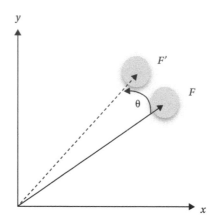

Fig. 3.1   Active transformation of the ball, about the origin, keeping the coordinate system fixed.

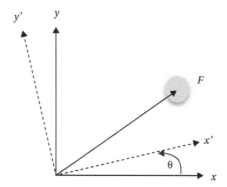

Fig. 3.2   Passive transformation of the ball, keeping the ball fixed but now shifting the coordinate system.

moving of the ball, from F to F' (by an amount θ) in fig. 3.1, all taking place in a fixed background coordinate system (x, y).

On a passive approach, the ball stays firmly where it is, but now we move our axes by the same amount θ, seen in fig. 3.2.

Being inverses of one another, combining the two transformations leaves us where we started (fig. 3.3).

Hence, it is a simple case of shifting the measured system versus shifting the coordinate system (the system of measurement). The next chapter puts the results from this and the earlier chapter to work in specific examples.

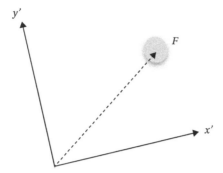

Fig. 3.3   Combining the active and passive transformations in the case of a symmetry leaves the original situation 'untouched.'

3.4 Further Readings

The philosophical discussion of symmetries naturally involves the usage of mathematics. The readings below give excellent guides to both the technical and conceptual elements.

Fun

- Hermann Weyl (1952) *Symmetry*. Princeton University Press.

    - Classic discussion of symmetry, which covers examples of symmetry in art, nature, and science and introduces with brilliant clarity the technical notions involved.

- Kristopher Tapp (2010) *Symmetry: A Mathematical Exploration*. Springer.

    - Though the title states that this is a "Mathematical Exploration," it is exceptionally easy to read, and will lead the reader to a good intuitive grasp of the relationship between group theory and symmetry.

Serious

- Elena Castellani and Katherine Brading, eds. (2003) *Symmetry in Physics: Philosophical Reflections*. Cambridge University Press.

    - Superb collection of readings, old and new, on philosophical aspects of symmetry.

Connoisseurs

- Eugene Wigner (1967) *Symmetries and Reflections*. Indiana University Press.

  - Much of our terminology and concepts come from this book. Written in crystal clear prose.

4 Getting Philosophy from Symmetry

Many of the central debates in the history of philosophy of physics concern the nature of space, time, and motion (kinematics) and geometric symmetries (symmetries of space, time, and motion). Lessons from these debates flow readily into a host of other areas, not directly related to spacetime. Space and time are, quite independently of tricky issues to do with motion and symmetry, rather strange. As Frank Arntzenius points out, we can't see them, nor smell them, hear them, taste them or touch them.[1] For this reason there is much overlapping between debates about space and time in the philosophy of physics and in philosophy more generally – more so than, e.g. the quantum theory of fields, for obvious reasons. We have a kind of direct access (Arntzenius' point aside) to space and time, that makes us feel like we know what we are talking about: space seems empty; time seems 'flowy.' But physics radically modifies our everyday views of space and time and so impacts on the more general philosophical debates.

This chapter builds on some of the core aspects of physics (kinematics, dynamics, states, observables, symmetries, etc.), developed in the previous chapters. We begin with the justly famous debate between Newton and Leibniz in which they agree about the world's overall geometry, but disagree over the ontological status of space and time: Newton will say that the world has its geometry because there is a space and time *with that geometry*; Leibniz will say that the geometry comes about from the laws linking material objects and events so that space and time emerge from this. Next comes a similar argument that links 'handedness' with the reality (or not) of space and time. After this we switch to relativistic physics: firstly, in special relativity, we focus on the twins paradox and a related argument about the reality of past, present, and future events; and then finally comes general relativity and the so-called 'hole argument.'

4.1 Leibniz Shifts and the Reality of Space and Time

I expect that for most people space is understood to be much like a really big room. Just as your bedroom contains your bed and wardrobe and so on, so space contains your bedroom, the Earth, galaxies, and all of the particles making up the universe's objects. Time is more difficult, perhaps, but again it is most likely to also be understood along 'container' lines in which events occupy specific instants and intervals of time, much as objects occupy the parts of space. As with ordinary containers, it seems easy to imagine space being emptied of its contents, and time emptied of its contents. But is this correct? Does a space emptied of all objects make sense? Is it a genuinely possible situation? What might an alternative look like? This is the debate over the ontological nature of space and time: are they *real* like chairs and hares and other material objects, or are they some kind of construct from chairs, hares, and such; or perhaps they are purely mind-dependent aspects of reality?

Absolute and Relative Motion

Space and time for Newton were certainly real. They were viewed to be quite independent from material objects, which they could indeed be said to *contain*. The motion of objects in this space then happened within this container, which is understood to have the same structure as Euclidean three-dimensional space. There is an 'absolute Cartesian reference frame' (x, y, z) in this space in which measurements must be made to formulate properly the laws of Newtonian mechanics – this analogy between a Cartesian coordinate system for Euclidean space and an (inertial) frame goes very deep, as we shall see. There would of course be motion of the objects *relative* to one another too, but *real* motion is understood to be motion relative to the special frame picked out by the absolute container. As we will see, this kind of motion was required in order to account for certain otherwise inexplicable 'inertial effects' (in which one must invoke a non-inertial frame). After all, in otherwise empty space, if we pass one another at uniform velocity in our cosmic armchairs we wouldn't be able to say which of us was *truly moving* and which at rest (or whether we were both in motion). But if you *felt* a force pushing you into your armchair, then we would know that you are truly in motion. For Newton, true motion simply meant absolute motion.

There is a problem, however: the special absolute frame is not unique. Rather, there is a special class of *frames*: the inertial ones. The laws (and the possible motions that satisfy them) do not depend on position, orientation, or velocity, which implies that given one 'special' frame, any other frame (axes: (x', y', z')) that differs from it by some rotation, displacement (translation in space), or even by some uniform velocity (a Galilean boost) will do just as well as far as the laws are concerned. This freedom in the choice of frame stems from a set of symmetries of Newtonian mechanics. The transformations that map some frame to another, in such a way that one still has a frame that is suitable for formulating the laws (i.e. an inertial frame), are symmetries of the theory: if some motion is a solution of the equations in one frame, then it will be in any frame that is related to it by applying one of these transformations. Of course, this means that one cannot use the laws to determine whether one is at rest relative to absolute space, or gliding against it at some uniform velocity.

This is all readily understandable if one assumes that space really is just like a Euclidean container. After all, the points and regions of the latter look the same regardless of the orientation or position of the axes. One is simply transferring the symmetries of Euclidean space (i.e. the distance preserving transformations of space, or *isometries*) to a physical context. But why believe in this absolute structure? Why believe in the reality of something that leads to such unobservable quantities as absolute position, orientation, and velocity?

Globes and Buckets

Newton believed he had *proven* the existence of absolute space by reference to certain physical examples in which relational structure would not be sufficient to account for some so-called *inertial effects* (or forces). These involve *rotations* (and accelerated motion in general, such as the cosmic armchair example above), but not understood as frames at different but stationary orientations: *actively* rotating systems. Rotations, he thought, must involve absolute motion. One of these examples involved a pair of identical globes connected by some cord, such that the entire system is rotating about a central axis. The rotation would, of course, result in tension in the cord so that it is pulled tight by the centrifugal force (see fig. 4.1).

How is the relationist supposed to deal with this apparently perfectly possible situation? The globes themselves are postulated to be identical. Given that they are the only things in an otherwise empty world, they also share all of their relational properties. So if there is tension in the cord the relationist has no purely relational resources to account for it. There must be rotation given the tension: the tension will give a means of measuring the rotation, and thus distinguishes the state of motion from one of uniform motion. Newton argues that the only way to account for it is to suppose that

Fig. 4.1   Newton's globes in an otherwise empty universe. The tension in the cord connecting them is an indicator of absolute rotation and a non-inertial frame.

they are rotating relative to the frame of reference provided by absolute space. In other words, motion cannot be merely relative in this case.

Another thought experiment, more famous than the globes, is 'Newton's bucket' from the Scholium to his *Principia Mathematica*. Here we are asked to consider a bucket filled with water suspended from some hanging point by rope. The idea is to consider what happens when we let the bucket spin. There are several stages through which the bucket/water system proceeds: first one imagines that the bucket has been twisted around many times so that there is potential energy in the rope that will spin the bucket when let go. What would we see? First the water would be flat and at rest, and the bucket would also be at rest. Then, as we let go of the bucket so that it can spin, the bucket will spin, but the water will remain flat and at rest. Then the bucket will be rotating and the water will be rotating, and will edge up the side of the bucket as it does so: again, giving an indication of absolute motion. Why *absolute* motion rather than merely relative motion? Because the bucket and the water are supposed to rotate at the same rate, and so will be at rest relative to one another. As Samuel Clarke (an ardent Newtonian put it, in his correspondence with Leibniz):

> [Newton] shows from real effects that there may be real motion in the absence of relative motion, and relative motion in the absence of real motion.

But there is a strategy for denying Newton's conclusion starting with Bishop Berkeley. Berkeley pointed out that if the spinning bucket, with water, were all that existed then it simply wouldn't make sense to speak of it as rotating: relative to what? The same point can be applied to the globes. The positivist physicist Ernst Mach, in his *The Science of Mechanics*, leveled a similar attack in the nineteenth century, pointing out that in the case of the bucket experiment, rotation relative to the mass of the Earth and the other celestial objects might be responsible for the non-flatness of the water's surface: the water is flat relative to these masses, and curved relative to these masses, rather than absolute space. In order to provide a decisive argument for absolute motion using the bucket (or the globes) we would have to empty the universe of all other matter. Though we might think we can do this in thought experiments, we can't rely on our intuitions in such cases: our intuitions might differ, but how are we to decide between them in such cases? The globes might expand or contract for all we know. However, Mach's response has its own difficulties: how, for example, do the masses of the various celestial objects in the universe act on our little bucket? They are very far away (some extremely far away), so does it take some time for the effect to occur? If not (if the effect is instantaneous, and so nonlocal), then surely this explanation is not as good as one that depends on local features of absolute space?

Ultimately, however, Mach was pointing out that since there is no way of distinguishing between a rotating bucket and the rotating heavens, the question of which is true is meaningless. That is, we have no way of assessing whether such inertial effects are the result of absolute motion or relative motion unless we could hold Newton's bucket still while spinning the rest of the universe around it, to see if that generated the same curvature in the water's surface by centrifugal forces. As Mach himself put it:

> The Universe is not twice given, with an Earth at rest and an Earth in motion; but only once, with its relative motions, alone determinable.

Whether one sides with the Mach–Berkeley strategy or not, it cannot be doubted that Newton does have an empirical argument on his side that needs to be responded to by the relationist.

In the age of airplane travel we can add another effect (that does not require rotation) observed during take off: the force felt that pushes you back in your seat is an inertial effect (this time generated by a linear acceleration, rather than the rotations of the bucket and globes experiments). You are at rest relative to the plane, and yet there is some force – here your body is much like the water sloshing in the bucket. Accelerated motion can be noticed: one can tell an accelerated frame from one at rest or in uniform motion. Just try and drink a cup of tea in an accelerated frame (a car taking a sharp corner), compared to a uniformly moving frame (a peaceful train journey). Whether it comes about through motion against an absolute container (i.e. departure from inertial, straight-line motion in that container) is the issue, however. The problem for the relationalist is that only one reference frame feels the effects: the runway accelerating from your plane at the same rate feels nothing! Mach would, of course, point out that it is acceleration relative to the rest of the mass in the universe such that if we could hold you and the airplane still and whoosh the rest of this mass away you might still feel the same force. Hence, there is absolute motion, only it is relative to this 'cosmic frame' rather than absolute space. What is missing, however, is a *theory* embodying this idea.

So-called Barbour–Bertotti (after Julian Barbour and Bruno Bertotti) models constitute a genuine attempt at a relationalist mechanics where the action takes place in a relative configuration space (essentially a space of all possible shapes, with no redundancy). One of its key Machian features is a constraint outlawing the rotation of the universe – or rather postulating a symmetry between the universe undergoing rotation, while an object (subsystem) stays fixed, and that subsystem rotating instead, with the rest of the universe as a whole remaining fixed. This is needed to secure a relationalist dynamics, so that fixing some initial relative configuration (just specifying the relative distances of the objects) will be enough to fix the relative motions forever after (since the whole system is in inertial motion)

– without this we would need to supply more data (an axis and rate of rotation) to get the subsequent evolution.[2]

Leibniz's Principles

Leibniz wanted nothing of this absolute container, which he believed allowed for a multiplicity of indistinguishable yet physically distinct possible worlds. To understand why Leibniz thought this was a bad thing, we need to quickly say something about his general philosophical principles: the principle of the identity of indiscernibles [PII] and the principle of sufficient reason [PSR]. The former principle has become an integral part of many issues in the philosophy of physics relating to symmetries. It simply says that objects sharing all of their properties in common are really just one and the same object, perhaps given different names or labels, which merely serve to (over-) represent the object. We can write this using logical notation as follows:

$$\forall F[Fx \equiv Fy] \to x = y \qquad (4.1)$$

In words, this says 'for any properties F, and objects x and y; if x has property F if and only if y has the property F, then x and y are identical.' Or, to put it another (converse) way: there is no distinction without a difference. One can also understand this principle to mean that there cannot exist two things that differ *only* in number, i.e. only in that there are two of them.

This sounds simple enough, but there are several subtleties involved in making sense of the principle. Firstly, there is an issue over what properties are to be included here: just qualitative ones? Spatial and temporal locations? There have even been proposed certain special non-qualitative properties known as 'haecceities' (primitive thisnesses). In ordinary language, a primitive thisness says that something is what it is *because it just is, so there*! 'Haecciety' is just a fancy word for this-ness. We can understand it as a certain kind of non-qualitative property (qualitative properties are 'suchnesses'), which involves being identical with a certain individual (obviously only possessed by a unique individual). Denying that such things exist, as a PII-wielding Leibnizian naturally would, commits one to a 'bundle' theory of individuals, according to which an individual's identity is just given by the various properties it has: there is no ultimate 'pincushion' sitting under all of the properties that gives each object its distinctive identity, so that even objects sharing every other property will differ in at least some way. Another issue is over *what* kinds of things are being compared: objects or possibilities? Or even entire worlds?

The PSR simply says that "for anything that is the case, there's a reason why it should be so rather than otherwise." If you have blue eyes, there has to be a reason for that: it can't be a whim of nature. Same goes for every single element and aspect of our world: including the way the material universe is configured and, if absolute space should exist, this also includes location within it.

The Shift Argument(s)

Leibniz used these principles to great effect in an argument that philosophers like to label the 'shift argument.' There are really several types: static, kinematic, and dynamical (depending on whether we use translations and rotations, boosts, or accelerations). Recall that the symmetries of Newtonian space mean that the laws cannot be used to distinguish between frames related by Galilean transformations. There is no mechanical experiment one could perform to tell which situation was the 'real' one – accelerations are a different kettle of fish, of course, as we will return to.

The Static Shift

Now, let us suppose that this Newtonian space is real, and the theory correctly describes our universe. Let's suppose that in this universe the tip of your nose is sitting at point x at some instant t. That means, given the symmetries of the theory, there is an infinitude of alternative possibilities (consistent with the theory) differing by some translation, or rotation, or boost, or some combination of them, leaving the tip of your nose at an entirely different (though no less real) point.

The problem (absurdity) that Leibniz draws out is that these possibilities would be qualitatively identical (indiscernible). The world in which the tip of your nose is at x and that in which it is at $x + 2$ feet westward (generated by rigidly shifting all of the matter in the universe by $x + 2$ feet westward: an element of the group Gal from the previous chapter) cannot be distinguished since all measuring devices have been shifted by the same amount: all relational material structure is preserved by the transformation (this is the meaning of isometry) – let's call these shifted worlds 'Galileomorphs.' We could alternatively leave your nose at x and shift the time at which your nose sits there, from t to $t + 2$ seconds (generated by another element of Gal). Again, there is no way to distinguish between t and $t + 2$ (remember, they smell, sound, taste, feel, and look the same!), so the worlds are strictly indistinguishable. Yet for Newton (and the believer in absolute space and time) they are physically distinct.

For Leibniz this is a massive violation of his principles. One can't have indistinguishable yet distinct entities. But, worse (as far as Leibniz was concerned), if it were true then it would mean that God created the universe when and where he did without having a good reason for doing so: not cool, says Leibniz. The argument can be 'de-theologized' quite straightforwardly. The fundamental point is that Newton's theory generates a bunch of solutions that cannot be physically distinguished in any way – we don't need these to fall under God's gaze for this to be the case. This is epistemologically unsatisfactory, since it means we cannot ever tell which of these solutions is the 'true' solution. If we think of possibilities as in some sense real things (possible worlds), then we have a further problem, which is that it is difficult to see in what sense these are genuinely distinct possible worlds. Leibniz's move, now called 'relationism,' was to cut out these indiscernible possibilities, collapsing them to one using his PII.[3] As he put it himself: "two states indiscernible from each other are the same state" and the idea that one could shift the entire contents of the universe is a mere "fiction."

All of the indiscernible worlds match up with respect to the material relations they embody: they are relationally identical. If one has an ontology of relations, then the worlds are viewed as one and the same: the apparent differences are not really physical. This is relationism: space and time are viewed as kinds of 'secondary qualities' that depend on distance and direction relations between material bodies and/or events. One also shrinks down the kinds of 'fundamental stuff' that exists in this way: there is only matter standing in various relationships, not matter *and* space and time (understood as equally fundamental) as in Newton's universe. Of course, if the multiplicity of indiscernible possibilities are not seen to be real, then absolute space cannot be real either. QED!

This pair of positions forms the basis of the modern debate over the ontological status of space and time, and has spread into other debates in which some symmetry is at the root of possibility generation: wherever there is a physical symmetry, there is the potential to generate indiscernible possibilities, which translate, in terms of physics, into empirically inaccessible structure. Whenever one has indiscernible possibilities one has an interpretive fork: treat the possibilities as distinct or as identical (so that, e.g. the differences are merely arbitrary features of the mathematical representation used). The former is generally termed a *substantivalist* approach and the latter a *relationist* approach – these terms stretch beyond positions concerning space and time.

There are problems with Leibniz's strategy of response. For example, Sklar ([46], p. 180) points out a way that the substantivalist might respond in a way that offers a sufficient reason: the material content is where it was yesterday and no force interfered with this during that time, so here it remains sitting perfectly consistent with Newton's laws. To be *otherwise* would violate the principle of sufficient reason. The problem with this is that it misunderstands Leibniz's point. Even given that it was in the same location yesterday and no forces shifted it, the question remains "why wasn't it at some other point yesterday and, given the absence of forces, today?" That is, why should it have spent its entire existence (even for all eternity) at some undistinguished point, given that all others would be just as suitable? Pointing to prior causes does not yield an answer to the issue of the *contingency* of the specific points at which the matter sits. Might there be another possible world, also in which we can say the same thing about its location (it's always been there, with no forces to disturb it), yet this location is distinct from the other in absolute space?

Or there might just be some irreducible randomness involved in the fact that since all points are the same any will do as well as another. Perhaps God threw a cosmic dart to find the central point, or did the trick where one closes one's eyes and simply drops a fingertip on a map to decide where to go, only here the decision was where to place the universe's contents. We are used to symmetries being broken 'spontaneously' in modern physics (though it is not without controversy); but the point is that we are less likely to be swayed by Leibniz's talk of 'God's sufficient reasons' for action.

The Kinematic Shift

Absolute locations might be passed off as detectable in a way (though not literally observable): objects are where they are in absolute space, period. We can't get a high-powered microscope and look at the spatial points, but if they are there then the body will be stationed over some particular points at any instant. Absolute velocities are more difficult since they require motion in a direction. Though we can say where we are in absolute space at any instant (namely right *here!*), we can't say where we are going and at what speed. The kinematic shift, involving Galilean boosts, causes similar problems to the static case, but there is a difference then: although there will indeed be infinitely many possible indiscernible solutions (being isometries), each with a different uniform velocity through absolute space, there is, in terms of the PSR at least, a reason to pick one as 'privileged' in some sense: the state of absolute rest has certain features that make it special and worth realizing if you are a creator that likes to act rationally. Because it does not involve any motion it has no direction that needs to be selected and so the isotropy of space needn't be a cause for concern. So while there is not a distinguished origin in terms of location, there is a kind of distinguished origin for velocity. But this leads to another shift problem.

The shift problems all spring from the independence of Newton's laws of motion (his equations) from time, position, velocity, and orientation, all of which spring themselves from the homogeneity and isotopy of space and time. Instead the equations depend on *relative* properties: configurations and velocities. But they *do* depend on accelerations. What we need is a way to distinguish accelerated motion from unaccelerated motion. One way to do this is with spacetime diagrams, in which lines (worldlines) represent trajectories of particles (see fig. 4.2): covering some spatial distance in some interval of

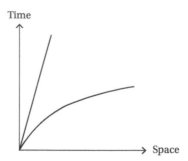

Fig. 4.2   Trajectory of two particles in absolute space (restricted to one spatial dimension and time): the straight line represents a particle with uniform speed while the curved line represents an accelerating particle.

time. The faster an object moves the smaller the angle between the worldline and the spatial axis: infinite speed would correspond to the worldline's being parallel with the spatial axis and rest corresponds to the worldline's being parallel to the time axis. Straight lines represent constant velocity. Therefore bent lines will indicate departure from constant velocity: acceleration!

The Galilean symmetry of Newton's laws translates into this spacetime diagram picture into an equivalence of tilted, yet still parallel worldlines generated by a Galilean boost. Hence, in fig. 4.3 (now with an additional dimension shown to reveal the different instants of time) the straight trajectories are indistinguishable yet distinct if absolute spacetime exists.

Hence, we have an awful lot of redundant structure in this representation. We have a unique way of splitting time into three-dimensional snapshots (Nows), corresponding to the unique, privileged inertial frame (with zero velocity), but the laws cannot determine which points of space the particles travel through on each slice. The kinematic shift simply employs the various families of parallel lines in place of the worlds shifted in terms of their absolute locations in the static case and Leibniz's two-pronged attacked, using his two principles, can get a purchase.

But all we really need to do justice to Newton's laws is a notion of when a line is straight and when it's bent (in the technical jargon, we need an 'affine structure') to distinguish inertial from non-inertial motion. This requires absolute acceleration and simultaneity, but reference frames (families of lines above) that differ by Galilean boosts are relative. So-called *Galilean spacetime* encodes this reduced structure, preventing the shift arguments from gaining a foothold. In effect, PII is being imposed on the spacetime but in a very restricted way (to absolute rest and velocity).

This eliminates a serious epistemological defect in Newton's version. In order to get a foothold there needs to be a notion of sameness of points from slice to slice (so that we can speak of different constant velocities), which Newton assumes, but this goes beyond the notion of straightness

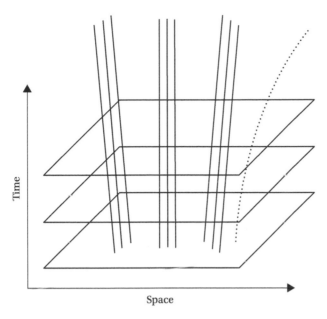

Time

Space

Fig. 4.3   Indistinguishable motions (straight lines) of particles in Newtonian spacetime representing physically distinct velocities, with the untilted lines in the middle representing particles at rest. The curved dotted line shows an accelerated particle once again.

of lines across slices. Our ability to detect absolute accelerations does not depend on this additional structure. Removing this renders the world and its boosted counterpart utterly indistinguishable (not just undetectable, as it is for Newton's spacetime). Galilean spacetime only cares about relative velocities, in keeping with the symmetry of Newton's laws, so that infinitely many states in Newton's framework correspond to a single state in this new and improved framework.[4]

The 'dynamical shift' is a different matter entirely, since it involves perfectly detectable forces. These give Newton's system its power. We turn to these issues next, but strictly speaking the notion of a shift argument in this sense threatens relationalism (along Leibnizian lines): a world and an accelerated counterpart are not indiscernible, so none of the usual tools for ridiculing absolute space (PII and PSR) are available. However, in purely relational terms they are indiscernible: accelerations are only relative motions. In the shifted world version of this, one world is at rest, say, and the other receives a global acceleration. Given the global nature of the transformation nothing relational would change, yet we might expect to feel a force in the accelerated case. It is possible to dig one's heels in and point out that we can't say what would happen in the global situation. Still, the law of inertia remains a problem for relationists.

Real Motion

We started with a discussion of motion, rather than the reality of space and time. It turns out that Leibniz agreed with Newton that certain states of motion (the non-inertial ones) are 'true (i.e. non-relative) motions,' but seemingly he doesn't see this as significant for the ontology of spacetime debate:

> I find nothing . . . that proves, or can prove, the reality of space in itself. However, I grant there is a difference between an absolute true motion of a body, and a mere relative change of its situation with respect to another body. For when the immediate cause of the change is in the body, that body is truly in motion; and then the situation of other bodies, with respect to it, will be changed consequently, though the cause of that change not be in them. (LV.53)

But how can this be squared with his overarching relationism about space and time (which must surely infect his view of motion, given that motion is couched in terms of space and time)? Newton had a response, of course. Indeed, his very reason for postulating absolute space and time was to provide a response to the reality of inertial effects (in which real forces are felt because of motion): absolute space is needed to make sense of inertial motion (undisturbed motion in a straight line); absolute time is needed to make sense of constant (i.e. unaccelerated) speed. If we can't ground inertial motion then how do we ground non-inertial motion? Leibniz grounds it in the bodies themselves, independently of relations to space and time, but also of other bodies. Real motion is that caused by force, and this, for Leibniz, is grounded in what he calls "*vis viva*" ('living force' or that which is responsible for force and so for real motion). This suggests that if we want to know which object is in true motion (and don't want to rely on the inertial effects themselves: looking for an explanation instead) then we must search for the causal origin of the motion – in the case of the airplane and the runway, we can see that the engines caused the motion. No mention of absolute space or time here. But there is no real story of how the idea works, and what it means to contain *vis viva*.

Similarly, Lawrence Sklar ([46], pp. 229–234) has argued that relationism and inertial effects can in fact be squared (and the debate about motion separated from the debate about the existence of space and time) if we take absolute states of motion to be brute, 'intrinsic' features of the objects that have them (not in need of deeper explanation), rather than being extrinsically related to absolute space and time (or even other objects, *à la* Berkeley and Mach) – monadic properties of objects rather than relations to spacetime. Hence, the relationist can believe that space and time are constructs from material objects, but add absolute acceleration as one of the possible intrinsic properties to be had by these objects.[5]

Mach's response is simply not to take the bait of these 'real motion' examples. They all rest on unverifiable thought experiments, as we saw above. But, as mentioned earlier, finding an actual theoretical scheme that can do everything Newton's absolutist scheme can do is a difficult challenge, the Barbour–Bertotti models notwithstanding.[6]

4.2 A Handy Argument for the Substantivalist?

The shift arguments in the preceding section made use of the isometries of Euclidean space (as well as Galilean boosts) to generate indiscernible possibilities. Another isometry is that of reflection. This cannot be achieved by a rigid motion (it is a *discrete* symmetry), so it is distinguished from the other motions accordingly. However, we can still draw similar philosophical consequences from it.

We tend to take the phenomenon of handedness (sometimes called 'enantiomorphy') for granted, since most of us are born with a pair of hands and feet. Our very DNA too comes in both left and right-handed flavours. Some fundamental processes in physics appear to favor a particular orientation: "God is a weak left-hander" as Wolfgang Pauli once said. But how is it the case that there are left and right hands? It seems to have something to do with space, but what precisely, if anything, does it tell us about the nature of space? Also, is the switching of the handedness of something a fundamental symmetry? How do we explain the difference between left and right? The situation poses formidable problems for those that wish to be relationists about space, for if handedness really is a spatial feature, then presumably, in principle, by their lights left and right hands are identical. After all, they can share all of their internal relational properties (same distance from base of thumb to knuckle of forefinger and so on) and yet they are evidently not quite the same, as can quickly be seen by trying to put a left-handed glove on right hand, to use an example due to Immanuel Kant.

Incongruent Counterparts

Kant repeatedly returned to the problem of left and right – 'incongruent counterparts' as he called left and right-handed entities: "[a]n object which is completely like and similar to another, although it cannot be included exactly within the same limits" – over several papers spanning more than a decade. So formidable was the problem, that Kant used it (initially) to argue for (Newtonian) absolute space (and against Leibnizian relationism) – he later viewed the same phenomenon as pointing toward a more relational view, as we will see. To see why the problem of incongruent counterparts is so tricky, let's consider a thought experiment.

Firstly, suppose that God (or the Flying Spaghetti Monster, if you are 'Pastafarianly' inclined) wanted to create a catalogue of all possible (instantaneous) relative configurations (relative distances) for the various objects (and their parts) in the world. This would say such things as $d(apple, orange) = 2\ cm$, $d(Earth, Mars) = 401,000,000\ km$, and so on. Now imagine that the full catalogue is made: essentially it should be able to function as a construction manual for (the spatial facts of) the world. Is that really enough to pin down all of the spatial facts of the world, such that any other world that was built following the specifications of this catalogue would be identical? Kant says *No*, because one has failed to specify how the various objects (the apples and oranges) are oriented relative to a global (worldwide) system of handedness.[7] One could imagine two of the Flying Spaghetti Monster's colleagues, each given a copy of the catalogue, constructing non-identical worlds from the same relative configurations that are mirror images of one another, in the sense that the one cannot be superimposed on the other. They would be what Kant called 'incongruent counterparts' (see fig. 4.4).

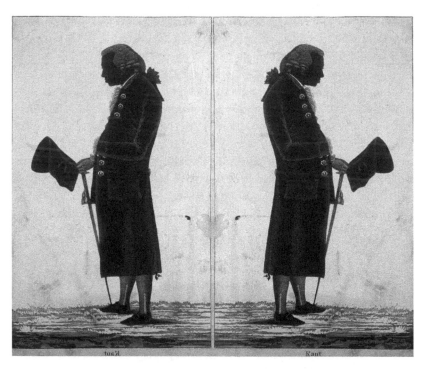

Fig. 4.4   Two possibilities constructed by following the 'relative-distance instruction manual.' Clearly distinct, since incongruent by any rigid motion, yet identical by the relationalist's lights.
[Image source: 'Immanuel Kant. Aquatint silhouette.' The Wellcome Library and The European Library, CC BY-NC]

This example clearly has a flavour of the Leibniz shift argument about it. We have a pair of worlds, that share all of their relational properties (under the reflection operation) and yet, there seems to be something genuinely different about them in this case (and *unlike* the case of the static shifts generated by translations). Yet Leibniz writes directly of this scenario that to ask why God created the cosmos as it is, rather than its mirror image, is to ask "a quite inadmissible question." He views the operations of (global) reflection and (global) translation as much the same: neither generates a discernible difference. We have been a little unfair on Leibniz since he would consider the entire universe reflected so that we don't have the luxury of comparing it with another, with both embedded in some larger space. Yet the relationalist is nonetheless left with the challenge of explaining *in virtue of what* a right hand is different from a left hand. So we have two issues: one concerning handedness in the context of whole possible worlds and one concerning handed objects within a world.

The Lone Hand

Kant presents the following thought experiment that appears to demonstrate that contrary to Leibniz, even the possible world version is an admissible question:

> Let it be imagined that the first created thing were a human hand, then it must necessarily be either a right hand or a left hand. In order to produce the one a different action of the creative cause is necessary from that, by means of which its counterpart could be produced. ([28], p. 42)

Hence, Leibniz had assumed that nothing could hang on simply reflecting everything in the same way: it would generate the same physical possibility and therefore is redundant. This redundancy would land God into a predicament in which there was no rational reason for actualizing one or the other. But Kant points out that handedness has a peculiar feature: there does seem to be an observable difference. If, for example, we assume that *another* entity is introduced into such a 'lone hand' world then it must *break* any Leibnizian-assumed symmetry. A hand would be shown to have been left or right all along, and hence there is a real difference between the left and right worlds. The alternative, for the relationalist, would have been that before the introduction of other handed objects, the handedness of the hand would have been indeterminate so that it would fit equally well on either side of a human body, which Kant views as absurd (it would surely only be congruent with one of the body's hands). Facts about the orientation of things must go beyond purely relative facts. Kant claims that there is some "inner difference" between objects of different handedness. Kant put this difference down to their absolute spatial properties. Hence, Kant argued that handedness reveals the existence of absolute space:

> [T]he determinations of space are not consequences of the situations of the parts of matter relative to each other; rather are the latter consequences of the former. It is also clear that in the constitution of bodies, differences, and real differences at that, can be found; and these differences are connected purely with *absolute and original space,* for it is only through it that the relation of physical things is possible. ([28], p. 43)

We are left with explaining in virtue of what left and right hands differ. Is it some internal, intrinsic feature they possess, or something external to them? Kant himself suggests that it involves reference to the space as a whole. If we have absolute space at our disposal then we can use the points of absolute space to point to some real difference between incongruent counterparts, even though it is a non-qualitative grounding.[8] But a simple passing of the buck to absolute space must do more than have a set of points underlying left and right hands; after all, the configuration of points will face the same problem: why are they left and right-handed? Hence Kant's reference to global properties of a space: it is a *relation* between the hand and (some structure of) absolute space that is supposed to do the work. This structure should allow for two ways to embed objects in the space – this is precisely why Hoefer believes that primitive identities are needed for the points in considering the two embeddings *as* two embeddings (see note 8).

Can Relationalists Handle Hands?

Is the situation for the relationalist as dire as Kant makes out? Or does the relationalist have some way of accounting for handedness in the world, and the role it plays? Recall that Kant's argument says that incongruent counterparts can't be explained relationally because there is simply no relational difference to be found in them. His lone hand scenario was supposed to clinch this. Yet the relationalist is capable of pointing out that handedness is an extrinsic property of objects: a lone hand has no external relations yet. In introducing, say, an opposite hand, then we have a pair of hands, but neither has its handedness *intrinsically.* Indeed, incongruent counterparts *are* intrinsically identical, but the reason for their incongruence comes from relations holding *between* them.

So a lone hand simply *isn't* left or right when alone in the universe: it has no handedness. Left picks out the class of things that have a family resemblance ('fits') with whatever was (conventionally) the first left. Nick Huggett ([26], §16.2) calls this the 'fitting account.' The idea is that 'congruence' is really an equivalence relation on the universe that partitions all of the objects it contains into equivalence classes {Leftys} and {Rightys} with 'left-handed' and 'right-handed' simply designating the members of these respective classes. One checks for the handedness of some particular object by checking the *fit* between it and the classes. Clearly we need more than a lone hand to do this, so Kant's premise is evaded. The problem was in supposing that handedness was an intrinsic property. Incongruence is then explained in terms of the *spatial relation* holding between them as material objects. (In the case of the Möbius strip world, in such a case the fitting account would be perfectly consistent with the idea that left and right handedness is a local property of the world, so that the partition into lefts and rights cannot be extended throughout the space. The substantivalist account might not fare as well in explaining left and right handedness in such a non-orientable world.)

Broken Mirrors

This highfalutin metaphysics can be linked rather directly with 'real physics' by simply asking whether the kinds of mirror reflections considered above are symmetries of physical processes and laws in the same way that, e.g. translations are. What this would mean is that there is no preference given by the laws of physics for some orientation. The laws would operate obliviously to switchings of left and right-handed versions of processes if this were so. It seems like a reasonable assumption, given the isotropy and homogeneity of space: why would physics care if we switched left to right? This was certainly the default position of most scientists in the first half of the twentieth century (until 1957: see below). For example, in his popular book on symmetry, the great mathematician-physicist-philosopher Hermann Weyl wrote:

> The net result is that in all physics nothing has shown up indicating an intrinsic difference of left and right. Just as all points and all directions in space are equivalent, so are left and right. Position, direction, left, and right are *relative* concepts. ([54], p. 20)

Weyl, by his own admission in later paragraphs, was following in the footsteps of Leibniz in saying this. Martin Gardner writing more forcefully (with reference to the Ozma problem of note 7) states:

> We are forced, therefore, to concede that our original problem is insoluble. There is neither a formal nor operational definition of left; no means by which it could be communicated to our sister planet. Another way of formulating this surprising conclusion is as follows: *Every known inorganic asymmetric structure or phenomenon exists in two mirror image forms identical in all respects except left-right orientations.* Mother Nature is ambidextrous. Apart from living organisms, she has no right or

left-handed habits; whatever she does asymmetrically, she does in mirror image forms. ([17], p. 210)

He rather presciently adds that "[t]here is no *a priori* reason why science might not tomorrow discover some type of structure or natural law which throughout the cosmos would invariably possess a left-handed twist" (ibid.).

Counterintuitively, this symmetry (parity symmetry) is indeed violated, for certain lawlike processes (namely, those involving the so-called weak interaction). What we find is that electrons (or beta particles) in a beta decay process will preferentially be shot out of the South side (relative to a strong magnetic field) than the North side. The original experiment was carried out with Cobalt-60 atoms. When such atoms are cooled close to absolute zero, the usually random scattering of electrons from the nucleus is focused into North and South channels. This setup clearly allows one to look at the rates of electrons going in both directions. A world with mirror symmetry would see no difference: why should the world prefer one direction in space than another? Surely the atom is much like Jean Buridan's donkey between a pair of identical bales of hay? But the electrons did prefer a direction in space allowing for a physical definition of a South Pole (that in which electron rates are highest).

To return to the 'Ozma problem,' we can now see how it might be possible to communicate what we mean by left and right, thus enabling any technologically advanced civilization to reproduce any pictures we might send in the right orientation. In six 'easy' steps: (1) get some Cobalt-60 atoms, (2) cool them near to absolute zero, (3) align the nuclei spins with a strong magnetic field, (4) count the emitted electrons, (5) call "south" the end with the most electrons emitted, (6) label the ends of the applied field accordingly, transfer these labels to the ends of a magnetic needle, position the needle over a wire in which current flows away from you: left is then where the north pole of the needle points.

John Earman argues that the existence of such a *lawlike* left–right (or parity) asymmetry (i.e. as opposed to the mere contingent existence of lefts and rights) makes life far more difficult for the relationalist interpretation of handedness. Indeed, he views the failure of mirror symmetry for the laws of physics as "an embarrassment for the relationist account"! As he explains:

> Putting some 20th century words into Kant's mouth, let it be imagined that the first created process is $\pi^{--} + p \rightarrow \Lambda_0 + K^0$, $\Lambda_0 \rightarrow \pi^{--} + p$. The absolutist has no problem in writing laws in which [one process] is more probable than [its mirror process], but the relationist . . . certainly does, since for him [they] are supposed to be merely different modes of presentation of the same relational model. Evidently, to accommodate the new physics,

relational models must be more variegated that initially thought. ([9],
p. 148)

That is, the usual 'Leibniz equivalence' manoeuvre (i.e. viewing situations
with no intrinsic differences as physically identical) simply fails here since
there are non-trivial differences. But there is nothing preventing the application of the 'fitting account' here. The "first process" of which Earman
speaks is no different to the lone hand of course, and we can say that without comparative processes it has no orientation. The relationalist can, then,
describe parity violating phenomena (a spatial asymmetry in the ejection
of electrons), and so encompass the laws of such processes, yet they do not
explain the asymmetry, treating it as a brute fact about reality. We might
have lingering 'principle of sufficient reasons'-based doubts about whether
this is good enough. As Carl Hoefer quite rightly points out, the hidden
assumption that makes the relationalist's response seem underwhelm-
ing is that the processes are taken to happen against a background space
that allows for multiple possibilities so that the nagging question "how do
those subsequent decaying pions know which direction is supposed to be
the more-probable one?" is faced (ibid., p. 252). It seems like a mystery
(pre-established harmony) how all of those electrons know to go south
given they don't have any intrinsic quality within them that makes it so. But
ultimate explanation is not on the table.[9]

4.3 Special Relativity: From Twins to the Block Universe

> There was a young lady named Bright,
> Whose speed was far faster than light;
> She started one day
> In a relative way,
> And returned on the previous night.
> [A. H. Reginald Buller in *Punch Magazine*, 1923].

The Classic Twins Paradox

The twins paradox of special relativity is one of the classic thought experiments in philosophy of physics. It appears to show that according to special
relativity, for a pair of twins, one of which undergoes a round-trip into
space at high speed, they will both appear to have aged less relative to the
other when they meet again. Hence the initial paradox: one cannot be both
older and younger simultaneously! The problem is, of course, that according to special relativity (in which only relative motions matter) either twin
can be considered to be the one that remains at rest (in the 'rest frame')
while the other dashes off. Though it might seem unnatural to suppose that
the spacebound twin is at rest while the other twin (along with the Earth!)

whooshes away, from the point of view of the physics, there is no absolute rest frame to ground the truth of one description over the other: all inertial reference frames (roughly, those in which Newton's first law holds) are equivalent from the point of view of describing physical processes.

But this apparent paradox is easily dissolved: relativistic time dilation *will* occur, because of the high speed of the journey, one needs to figure out what feature is responsible for the decreased ageing of the space-traveling twin rather than her Earthbound counterpart: what is the nature of the asymmetry? The solution lies in the fact that only one twin will complete a journey in which there is a 'turnaround' to make the return journey. Perhaps they slingshot around a star or hit the reverse thrusters. This simple fact means that the spacebound twin must occupy *multiple frames of reference* (i.e. they will not be in an inertial frame, characterized by constant velocity, for the entire journey), while the Earthbound twin stays in a single inertial frame (since the Earth is in free fall). In which case the spacebound twin indeed ages less. The symmetry that would otherwise allow us freely to use either description (spacebound twin at rest or in motion) is therefore broken, since that only holds for inertial frames. But we still need to say exactly why the traveller ages less, and what changing frames has to do with it.

This solution makes the problem look rather trivial, and you're perhaps wondering why I referred to it as "a classic." However, there are still interesting features to probe, including some that have only recently emerged in which the turnaround manoeuvre is removed by a clever topological trick. This 'topological twins' scenario allows us to dispose of a common answer to the question of what causes the difference in ageing: *accelerations* (or the physical nature of the turnaround process itself) during the switch from an outbound to an inbound trajectory – the latter is closer to the truth, but still isn't quite the proper explanation. Let us develop some of the details of the twin paradox setup.

Firstly, note that the choice of identical twins is simply done to make the example more colorful: all that matters is the *differential ageing* that results from the high-speed (relativistic) travel. Let's name our travellers Angelina (Jolie) and Brad (Pitt). We can simply have them wear twin watches if we wish to, to inspect the difference in seconds of proper time passed. (The time shown on their watches (or in their biological processes, which also function as a clock of sorts) is known as the 'proper time' and depends on the state of motion of the clock.) These watches will be synchronized before they split at $t = 0$ and compared when they meet again. Let us suppose that Angelina travels 4.22 light years away, to Proxima Centauri, at relativistic speed (forget about the biological implications of a human traveling at close to the speed of light). Their respective spacetime trajectories (viewed from Brad's frame) would look as in figure 4.5. According to this diagram, of course, it looks like Angelina's journey is by far the longer. However, we

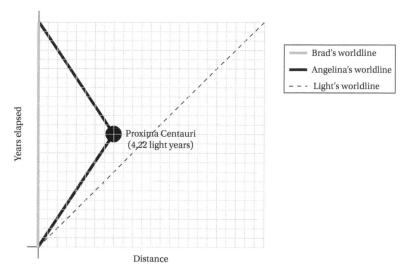

Fig. 4.5   Brad (light gray) and Angelina's (dark) trajectories in spacetime. The motion is plotted from Brad's perspective who simply 'stands still' traveling up the time axis, while Angelina whizzes off to the nearest star, and immediately returns (that is, we have assumed an instantaneous turnaround: we could smooth this vertex off adding additional time to the journey). Note that the 45-degree angle for light implies that it travels one unit of distance per unit of time (formally, we say that $c = 1$, where c is the velocity of light, since velocity is simply distance divided by time).

need to remember that this is simply a representation of the Lorentzian spacetime interval on a Euclidean page: the longer the spacetime interval, the shorter the journey in special relativity. The crucial element in the twins paradox is the dilation (or 'gamma') factor:

$$\gamma = (1 - v^2)^{-\frac{1}{2}} \qquad (4.2)$$

This factor gives the ratio for the relative rates of Brad and Angelina's wrist-watches (according to which Angelina's watch appears to run slow relative to Brad's). Note that it is entirely velocity dependent, with no sign of rates of change of velocity: the faster one travels, the greater the dilation of one twin's tick rate relative to the other – but note, in relation to the 'acceleration solution' above, that no mention is made in this factor of accelerations: dilations do not care about accelerations! Or they do only inasmuch as accelerations are implicated in speed changes. Note also, that special relativity is perfectly equipped to deal with accelerations, which would simply be represented using curved worldlines on a spacetime diagram.

Suppose we have mastered spaceflight to such an extent that we can instantaneously accelerate Angelina to 80% of the speed of light c: $v = 0.80c$. Let's round the distance d to 4 light years for simplicity. Brad would calculate Angelina's roundtrip to be just twice the distance to Proxima Centauri divided by her speed:

$$t_{Brad}(\text{round trip}) = \frac{2(\text{distance})}{\text{Angelina's velocity}} = \frac{8 \text{ light years}}{0.8c} = 10 \text{ years} \quad (4.3)$$

This simply means that Brad will be ten years older when Angelina returns home than when she set off. But taking into account the relativistic speed of Angelina, we need to include the γ-factor, $\sqrt{1 - \frac{v^2}{c^2}}$. Recall that $v = 0.8c$ and c itself is just 1 (the speed limit). So we have:

$$\gamma = \sqrt{1 - \frac{(0.8)^2}{1^2}} = \sqrt{1 - \frac{0.64}{1}} = \sqrt{1 - 0.64} = \sqrt{0.36} = 0.6 \quad (4.4)$$

So to find Angelina's age, Brad's ten years will be dilated by this specific γ-factor yielding $\gamma \times 10 = 0.6 \times 10$. Angelina will have only aged six years compared to Brad's ten (or, there are almost two ticks of Brad's watch for each tick of Angelina's). Remember, also, that since Angelina is in motion at high speed, her spaceship will be contracted in the direction of motion by the γ-factor, so that from her frame (in which she is at rest, of course) she will have covered only $0.6 \times 4 = 2.4$ light years, which explains the six-year-long round trip: $2.4/0.8 = 3$ (using $d/v = t$) for each leg – note that during the outward leg, the symmetry of their perspectives is preserved: either could speak of the other as the twin that moved. The 6:10 year ratio is a direct consequence of the dilation factor of 0.6 – as an exercise, try playing around with different values of the dilation factor in order to see how big an age difference one can engineer. This is not just a case of Angelina's watch showing that six years have elapsed rather than ten: she will have biologically aged six years, unlike Brad and the rest of Earth's inhabitants who have aged ten.

The impact of frame changing on the age difference can be seen with the aid of a spacetime diagram this time highlighting Angelina's simultaneity slices (see fig. 4.6):

One can see that the instantaneous switch in Angelina's direction of motion results in a chunk of Brad's time (four years' worth) being leapt over in terms of Angelina's notion of what is happening now. She will compute startlingly different results for Brad's age immediately before and immediately after her turnaround because of this frame-change. It should be clear than this is not based in the acceleration felt during the turnaround, and we have mentioned nothing other than plain vanilla special relativity. There is also nothing particularly mysterious about the time being leapt over: it is not 'missing time,' and is more of an artefact of the way Angelina's frame (and so her measurements) must alter with respect to her motion.

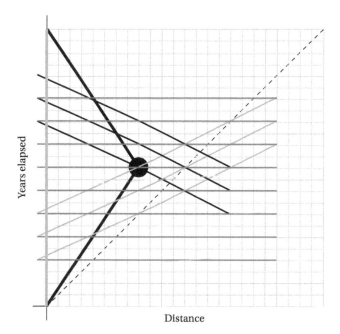

Fig. 4.6   As Angelina travels away from Brad, her simultaneity slices (what she considers to be happening *right now*) tilt relative to Brad's (which are simply the slices orthogonal to his wordline) in order to preserve the constancy of the speed of light (which, you will recall, we think of as covering one unit of distance in one unit of time) – Angelina's slices are also orthogonal to her worldline, as determined by the (Lorentzian) inner product associated with Minkowski spacetime. She turns around at the star, there is a sudden switch from one inertial frame to another, which results in the simultaneity slices tilting the other way.

There is a sense in which Angelina has performed a certain kind of time travel into the future: she has slowed down her own ageing (a kind of motion based cryogenics) so that she is out of phase with the ageing of those on Earth. Had she traveled faster and longer she could have returned tens of thousands of years into the Earth's future (relative to when she left) without ageing much at all – perhaps she finds the Earth scorched thanks to global warming? Of course, this is a one-way journey: there's no going back to Brad to let him and the rest of Earth know their fate. Any other round trips will only send her further into the Earth's future.

Topological Twins

Recent work on the twins paradox, and the question of what it is actually showing us about spacetime and relativistic motion, has focused on ingenious topological versions, for closed, non-simply connected spaces

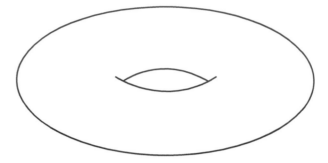

Fig. 4.7   A torus (or 'doughnut') characterized by a multiply connected topology. A perfect world for considering the twins scenario without the bother of the traveling twin having to turn around.

(i.e. in which parts of the space are identified, such as gluing the ends of a strip together to create a closed loop). For example, by confining Brad and Angelina to a closed cylindrical universe we can have Angelina complete her round trip without having to accelerate or turn around by simply completing a circuit around the cylinder. (We are assuming that it is only the surfaces that are relevant here, so that the setup would be something like the old-fashioned Asteroids computer game in which one leaves one side of the screen only to emerge on the opposite side. But the example can be generalized to a three-dimensional version, which would simply amount to walking through, e.g. your living room wall and coming out on the opposite side of the interior wall. Identifying one side of the screen with the other results in a non-simply connected topology.) Bear in mind firstly that there is no 'real' (intrinsic) curvature in this space: it is locally flat and can be constructed by taking a rectangular section of ordinary flat spacetime (the 'fundamental domain') and rolling it into a tube by identifying two sides, just as one can make a pea shooter by rolling up a page of paper. If we identify the remaining open ends then we will have made a torus (see fig. 4.7): simply imagine first creating the cylinder and then gluing the two ends of the cylinder together.

Poincaré-disc considerations aside (on which, see the next chapter), one could only verify that one lived in a cylindrical or toroidal universe by determining the global structure of the space, for example by sticking a marker into the ground and traveling in a straight line for long enough until one intersected it again. In this way one finds the 'loops' in the space. We can imagine a surveyor letting out a reel of string as they go, eventually coming back to their starting point, and we can imagine them tying a knot in the loop and pulling it tight. This would correspond to one of the 'cycles' of the space, of which there are two in this case: one around the handle and one around the hole.[10] The question is: will there be the symmetry in this space that led to the original twins paradox, so that each twin views the other's journey as longer?

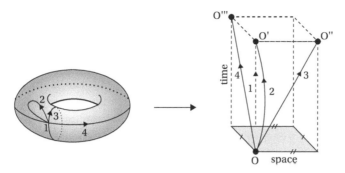

Fig. 4.8   The possible motions of quadruplets in a toroidal universe in which they must travel. The left side shows the paths they will take around the space (with 1 staying put), while the right side show their worldlines, beginning and ending at the same spatial location. The worldline 1 represents the stay-at-home quad. Image source: [31], p. 540.

There are various ways one could travel on the surface of this doughnut some of which have no counterpart in flat space. The crucial feature is the hole, responsible for a multiply connected topology involving a pair of non-contractible loops: there are two directions (around the hole and around the handle) in which one is prevented from shrinking loops down to points, as with the reel of string above. This is highly significant in the topological twins scenario. In fact, in a very clear presentation of this example, Jean-Pierre Luminet [31] considers a 'quadruplet paradox' instead, with each quad performing a different kind of motion in the space (fig. 4.8).

The quad that travels along the second worldline is simply doing the classic twins journey discussed above: a round trip with turnaround. We know that they will have a longer proper time than the stay-at-home quad and the symmetry between their perspectives about 'who was really in motion' will be broken because of the second quad's switching between different inertial frames. But what of the other two quads: they travel in perfectly straight lines (inertial frames) and at no point do they turna-round or initiate thrusters, availing themselves instead of the wraparound topology. If the symmetry is broken in this case, what breaks it? It cannot be a *change* of frames since they move inertially throughout their jour-neys. While not a change of frames as such, there is a difference in their frames caused by the non-simply connected topology.

The difference emerges when we inspect the so-called homotopy classes of the various journeys in the space which encode features of the space's global topology. We say that a pair of loops belongs to the same homotopy class (or are *homotopic*) if one can be morphed into the other by continu-ous deformation (i.e. without snapping or gluing either of them). The first and second quads' journeys are homotopic since the second quad loop can simply be shrunk to a point without meeting an obstruction (such as a hole or a rolled up dimension, again as happened with the reel of string) – they share the same winding index (0, 0). Their symmetry is broken (and the paradox resolved) in the standard way, by the existence of frame changes. The trajectories of quads 3 and 4, however, do involve the handle and hole of the torus and this means that their loops are mutually non-homotopic: they cannot be morphed into quad 1 and quad 2's trajectories, nor can they be morphed into each other – they lie in entirely different homotopy classes respectively characterized by the winding numbers (0, 1) and (1, 0).

So here is an asymmetry: there are non-trivial (topological) differences in the quads' trajectories. This causes differences in their frames and is already enough to dissolve the paradox. However, we are unable to see information about proper time. These windings can be visualized in what is called the *universal covering space* of the fundamental group, which has the effect of 'unwrapping' the loops wrapping around the torus. This simply means that we take the rectangle of flat space that we started with when we built our torus, and rather than rolling it up we simply tile the plane with it, but remembering where the original copy was positioned since this will correspond to our (0, 0) case in which no wrapping around occurs (see fig. 4.9).

One can now use the covering space to retrieve information about the proper times elapsed for each quad since this description includes metrical information. As before, the length of the worldline and the proper time are in inverse proportion: the longer the worldline the shorter the journey. In which case, quad 4 (with winding number (1, 0)) ages the least, followed by

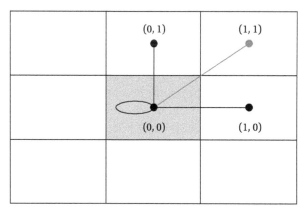

Fig. 4.9   Unwrapping the loops using the universal covering space for the torus, showing our quads' paths. Neighboring horizontal and vertical cells correspond to one winding, around the hole and handle respectively. Diagonal cells (1, 1) correspond to single windings around *both* handle and hole. The starting point and the end point correspond to the same point in space. To consider several windings one would have to move further out to other cells. One can see by direct inspection that the worldlines have different lengths.

quad 3 (with winding number $(0, 1)$), quad 2 (with winding number $(0, 0)$, but with accelerations), and then the poor stay-at-home quad 1 (also $(0, 0)$, but occupying the same inertial frame throughout).

Relativistic Reality and the Open Future

The twins paradox was found to be no such thing, no paradox at all: simply a feature of special relativity. It exposes a peculiar yet physically verified aspect of our universe: the faster you move the more slowly you age. Before we leave special relativity, let us consider how the relativity of simultaneity (found to be at the root of the twins paradox) has also been invoked to argue for some very deep metaphysical theses about the nature of reality. Time dilation and spatial contraction play no direct role here, and only the velocity-dependent tilting of worldlines is needed (though of course this implies dilation and contraction).

Recall that the relativity of simultaneity is the idea that the present moment ('the Now') is relativized to the state of motion of an observer, so that there is no unique such Now and observers that are in motion relative to one another will identify a different set of events as constituting their present moment – this is a three-dimensional spatial snapshot of the universe (the universe at an instant: a notion that will differ depending on an observer's state of motion) known in the literature as a *spatial hypersurface*. In a paper that sparked many responses, philosopher Hilary Putnam [38] argued that since it is possible to find pairs of observers such that present events for one, say Angelina, are to the future of the set of present events for the other, Brad, it must follow that those future events are *real* for Brad and so pre-determined (given that Brad is real to Angelina). In other words: according to special relativity the future is not *open*.

With the machinery of spacetime diagrams to hand it is simple to see how this 'fatalist' conclusion is supposed to come about – fatalism is the view that all events are predetermined: there is no contingency in what will happen. Consider the following diagram (fig. 4.10), modified from the twins paradox diagram above to show the planes of simultaneity for the observers. Thanks to the relativity of simultaneity, the great spatial distance between Brad and Angelina allows for the possibility of differences in their determinations of temporal separations between events (i.e. what they deem 'simultaneous,' 'before,' and 'after'). We don't need spatial separation to get differences in what events are considered to be simultaneous: one can simply have Brad and Angelina pass each other in opposite directions, perhaps as they manage a quick fleeting kiss, so that their simultaneity surfaces are not parallel. In this case, we can say at the instant of the kiss that they are both real (i.e. both exist in a determinate sense) for each other.

All of this thus far simply reveals just how different spacetime in special

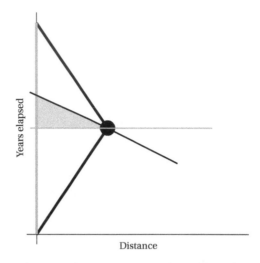

Fig. 4.10   As Angelina starts her return journey, her surface of simultaneity can be seen to contain events that are in Brad's future, as well as events to her past that are also to his future, shown here by the shaded area. This might include such events as Brad's winning the lottery. Likewise, there are some events on and to the past of Brad's surface of simultaneity that lie to the future of Angelina's surface of simultaneity. This will still be the case at the moment of contact if Brad and Angelina intersect as they travel in opposite directions. (Note that the same can be said of Angelina's outward journey, though with a different set of events that are to the future of Brad's surface of simultaneity so that what is to the future for Brad on Angelina's outward journey are to the past for Brad on her return journey (and vice versa).)

relativity is from earlier Newtonian physics. There we had a single Now dividing the events up neatly into past, present, and future. Here we have a more complicated affair, but we can still partition the events in the world according to how they can (or cannot) be causally connected by light signals or signals traveling slower than light. This classification of all of the world's events involves the spacetime interval built from the separate temporal and spatial intervals using the rule:

$$(\text{spacetime interval})^2 = (\text{time separation})^2 - (\text{space separation})^2$$

In Euclidean space, like that used in Newtonian physics, one can only ever speak of positive or zero intervals, since we only ever use *sums* rather than differences. The revolutionary aspect of special relativity is that we must introduce a third possibility: negative intervals. Hence, we get the following three ways that events can be related:

Timelike: (time separation) > (space separation)
Spacelike: (space separation) > (time separation)
Lightlike: (time separation) = (space separation)

Again, in Euclidean space a zero interval would suggest something uninteresting, pointing to the fact that two events are at the same place. In the context of special relativity we find that vast spatial distances can be linked by null spacetime separation so long as they are linked by light rays. The trick is to modify temporal measurements accordingly so that there is no time interval whatsoever for anything moving at light speed! In terms of the spacetime diagrams, consider how the simultaneity surface must tilt if one travels at light speed: it must lie parallel with the worldline (something enforced by the inner product). A beam of light does not experience events as separated in spacetime since its time and space separation will cancel each other out, hence the expression "null" interval. If it were possible to accelerate a spaceship up to the speed of light and run the twin paradox scenario again, then Angelina's watch would show that zero seconds have passed – i.e. her proper time would be zero and she would not have aged at all.

One can link this to Putnam's argument by noting that the events that are 'simultaneously' both past and future (past for Angelina; future for Brad) cannot be linked to Angelina by light rays nor any signal traveling more slowly than light: they are spacelike separated from Angelina (though timeline separated for Brad). Let us lay out Putnam's argument more explicitly, before considering the responses.

The argument invokes a link between what is real and what is in one's 'present snapshot' and also an assumption that this reality is *transitive*: if we are both real, then what is real to you is also real to me – this latter claim he calls the principle of "no privileged observers." With this in mind, we have (switching to our characters, Brad and Angelina):

1 Angelina-now is real.
2 At least one other observer is real and can be in motion relative to Angelina (that's Brad, of course).
3 If all and only things that stand in relation R [simultaneity] to Angelina-now are real, and Brad-now is also real, then all and only things that stand in relation R to Brad-now are real.

From this simple set of premises, Putnam concludes that according to special relativity "future things are already real!" That is, even though Angelina cannot communicate or interact with such events about Brad's future, the fact that Brad is another member of reality, combined with facts about relative surfaces of simultaneity of those in relative motion (combined with a philosophical assumption about simultaneity grounding what is real), it follows that Brad's future has 'already happened' for Angelina, and so for Brad too (by transitivity).

The ultimate conclusion of this kind of thinking is that the notion of a special three-dimensional surface (a Newtonian Now) carving out and

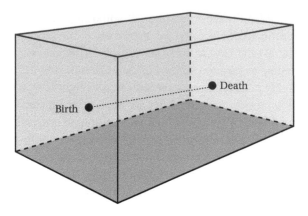

Fig. 4.11   The block universe picture of the world. All events in the block (the universe as a four-dimensional entity) are taken to have the same ontological status. At the instant of your birth your death is already etched into the block – and indeed, your birth too is etched into the block for all eternity!

constantly reshaping the past and future[11] to generate a four-dimensional universe cannot be sustained given special relativity with its multiplicity of Nows. Instead, it must be replaced with a single spatiotemporal 'block' (see fig. 4.11).[12] Thinking in terms of space and time as separate entities trips us up, and must be replaced with a fully spatiotemporal picture. But, the argument goes, this must mean that 'becoming' (in which reality has a dynamical character) has no place in modern physics.

The long and short of it is this: if there is no home for a present moment in special relativity, then there is no home for 'becoming' (nor becoming *real*) since that requires a division into 'past,' 'present,' and 'future' and any such division is frame-dependent in special relativity. Reality can't be frame-dependent: who's frame? All frames? So becoming must go and, many argue, the block must replace it, and the open future is closed up.

In the case of special relativity the disagreement in measurements only occurs if we take time and space as separate entities. If the focus is on the *spacetime interval*, there is no disagreement about 'physical facts.'[13] As Minkowski famously said in a public lecture of 1906:

> Space by itself and time by itself are doomed to fade away into mere shadows, and only some kind of union between the two can preserve their independent reality.

However, we must handle Putnam's argument with some care. We already saw that the events in question lie outside of Angelina's light cone and, therefore, can be made 'future' or 'past' by finding the right inertial reference frame moving relative to her (where we are now defining her by the separation point between her future and past light cones). If we have Brad and Angelina intersect as they move in opposite directions, then the events in question are outside the light cones of both, and so could never have anything to do with either. That such events exterior to the light cone are often bundled together and called "the absolute elsewhere" is no accident. There is much conceptual confusion over how to interpret the status of such events. For example, Howard Stein [49] has argued that such events cannot be said to be real with any justification in special relativity; rather, what is *real* (e.g. when Angelina turns around at Proxima Centauri) is what lies "at points in the topological closure of [her] past [light cone]" and this will be dependent on the origin of the spacetime point in question.[14] Stein puts it as follows:

> in Einstein-Minkowski space-time an event's present is constituted by itself alone. In this theory, therefore, the present tense can never be applied correctly to "foreign" objects. This is at bottom a consequence (and a fairly obvious one) of our adopting relativistically invariant language – since, as we know, there is no relativistically invariant notion of simultaneity. ([48], p. 15)

But Putnam considers this to constitute a kind of solipsism. only what has happened in Angelina's past light cone is real! What about the rest of reality? She was OK in *Tomb Raider*, but to claim that she is in charge of reality is going a bit far. Stein agrees that it is solipsistic, but it is so in a "pluralistic" way (ibid., p. 18), applying to any point: we must view what events have 'become' from the standpoint of some particular event rather than in a global fashion.

Again, and this is part of Stein's message, special relativity is a theory of space-time, rather than space and time. It is a theory of light cones constraining causal influence.[15] The whole setup of Putnam's argument involves an older way of thinking, in terms of shared present moments. Stein's ultimate target is with the view that special relativity necessarily involves a conflict with what philosophers call 'becoming,' where events are *not* to be treated as possessing equivalent ontological status, but depend for their reality on whether they are past, present, or future – becoming is, then, the idea that events 'become more and more past.' But time is multifaceted in special relativity. As Steven Savitt [42] clearly notes, there is coordinate time and proper (worldline) time, and the latter is perfectly well-suited for linking to a notion of becoming – this view was developed by Rob Clifton and Mark Hogarth [5]. What alters, however, is that becoming is localized to individual worldlines (observers), rather than to a global Newtonian Now. This might be hard to swallow: is this what we mean by the notion of becoming, of a dynamical conception of time? In a different way, Stein also argues that so long as we are willing to make some revisions appropriate to the shift brought about in the transition from

Newtonian to relativistic physics, one can make some (limited) sense of becoming in the world. We were perfectly content to align our notions of reality with the notion of a Newtonian present, and we usually (though often reluctantly) let our philosophical views about reality march in step with advances in physics. So, why not let relativity guide us here? If the present moment ceases to be an invariant notion, then, if we are realists about our scientific theories, our conception of reality must shift accordingly.

Though this is really a special relativity section, there is an interesting supplement to this issue that involves the theory of *general* relativity. An argument due to the great logician (and Einstein's friend in his later years) Kurt Gödel claims to show that there is (in at least one world that is possible in general relativity) no objective lapse of time [20]. In other words, no becoming: no situation in the world in which it would be true to say "the future is not yet determinate." What Gödel showed was that there could be physically possible situations in which there was no way to establish a global Now that definitively split the universe up into past, present, and future events. Without such a notion, he argued, one could not speak of objective *change* either, since change requires the lapse of time (one thing becoming another thing, changing color and so on). For this reason Gödel viewed time (and its related concepts) as entirely subjective or 'ideal.' We return to Gödel's universe again in the final chapter, where we consider its bearing on the possibility of time travel and time machines.

What we should draw from all of this is that the question of whether there is a present (and so becoming) is dependent on the physical conditions of the world: it is a matter for physics rather than philosophy alone to decide. Future advances in quantum gravity, for example, will no doubt serve to refashion (and perhaps reinvigorate) the debate.

4.4 General Relativity and the Hole Argument

Just as special relativity's symmetry of Lorentz invariance was at the root of the twins paradox, so the characteristic symmetry of general relativity, diffeomorphism invariance, lies at the heart of our next philosophical problem, known as the hole argument. Just as the twins were not really necessary for the twins paradox, so holes are not really necessary for the hole argument (at least not in its modern guise)! Rather surprisingly, the argument was originally developed by Einstein as a way of showing that one could not have a generally covariant theory of gravity since it would clash with Mach's principle. The details are a little convoluted, but basically the idea is this: Einstein believed that the matter distribution (i.e. the configuration of mass-energy in the universe) should determine a *unique* metric for spacetime. However, with generally covariant equations we have the freedom to alter the metric in various ways (using diffeomorphisms, which alter the metric smoothly, leaving 'deeper' aspects known as topological structure fixed) without leading ourselves from a possible solution (i.e. a matter distribution with a spacetime geometry: $\langle \mathcal{M}, g, \mathcal{T} \rangle$) to an impossible solution. That means that we can generate multiple (infinitely many in fact) solutions for the same matter distribution, in violation of Einstein's understanding of Mach's principle (as a broadly relational principle involving the idea that the rest of the matter in the universe determines the motion of bodies in even small regions: distant matter has an effect on local motions: inertial motion is governed by the aggregate of masses in the universe as opposed to a Newtonian container). The hole appears since the way the argument was originally set up involved altering the metric only inside a small hole in spacetime (where the hole is defined by the vanishing of matter within it: i.e. $T = 0$), around which the metric was fixed to a specific value. In this scenario, we can have knowledge of the geometry and matter outside of the hole (and on its boundary), but even with this complete knowledge, we cannot determine uniquely how the metric will develop into the hole: a failure of determination or causality since we can construct a coordinate transformation (a diffeomorphism) that only acts non-trivially within the hole – the modern version of the hole argument to be discussed in a moment simply turns Einstein's argument violating causality into a temporal one violating determinism, by essentially making the 'hole' the entire future to a slice through spacetime and adding the metaphysical component of belief in the reality of spacetime points (substantivalism).[16]

What Einstein wanted from his theory was that the geometrical features of spacetime were uniquely determined by the distribution of matter and energy. Before we get to this, and the hole argument itself, we should first briefly explain what general relativity is and how it works. The theory is rooted in the idea that spacetime (the history of the universe) is modeled by a four-dimensional manifold (think of this as a space that can be labeled by coordinates) equipped with a (Lorentzian) metric that specifies distances and angles between points of spacetime (i.e. events). The crucial difference with respect to all other spacetime theories that came before is that this metric obeys equations of motion (Einstein's field equations): it is a dynamical actor in the theory that couples to the state of matter and energy. The metric in general relativity multitasks, representing both gravitational as well as the above spatiotemporal features. This means that if mass and energy can act as a source of gravity (which they can of course), then they can also act as a source of warping of the geometry of spacetime – this was Einstein's understanding of the equivalence principle, which sits at the heart of general relativity: in terms of observable properties, a gravitational field applied to a reference frame is identical to an acceleration of the reference frame in the opposite direction. This was argued for using the famous 'elevator experiment' (a thought experiment, very similar to Galileo's ship example: I expect that Einstein had this in mind). Suppose you are confined in an elevator (with no way of seeing out) on the surface of the Earth, which has a mass that induces a gravitational acceleration on objects of 9.81 meters per second squared. Now suppose that an evil scientist floods the elevator with a gas that sends you to sleep, and then shifts the elevator into deep space, but straps a rocket onto the underside that accelerates you at 9.81 meters per second squared: you would not be able to tell by performing experiments located within the elevator that you had been moved at all. But, light beams allowed to stream through the elevator would appear to have a slight curve due to the motion of the elevator through space. This feature allowed Einstein to make the prediction that the gravitational field of a massive body must cause the light to curve in an identical way: this was tested (and confirmed) by Sir Arthur Eddington, who measured the deflection of light by the Sun during an eclipse. Since light travels along geodesics (shortest time/energy paths), it must be the spacetime geometry that is being 'bent' by the Sun (the gravitational source).

General relativity demanded a very high degree of symmetry (diffeomorphism symmetry) to perform its function.[17] In a Galilean invariant theory one mustn't be able to detect operations that translate, rotate, or give uniform boosts to the reference frame you are conducting your experiments in. Confined to your ship's cabin, with only Newton's laws, you shouldn't be able to figure out how the boat is moving and where and when it is. In the case of general relativity, the operations are generalized to any motions, including accelerated ones: in your spaceship's cabin, armed with the laws of general relativity, you can't tell whether you are accelerating or sitting on a planet by measurements using rods and clocks – in other words: there is no way of telling that an isometry (a spacetime distance preserving map) has been applied.

In more visual terms, this is the meaning of diffeomorphism invariance: warp your spacetime geometry from a perfect sphere into a teddy bear shape and the laws of the theory won't bat an eyelid. They won't notice since from their perspective all that matters are the topological properties (the invariants) and these don't care about stretching and squishing, so long as one doesn't tear the spacetime or glue pieces together (as with the cylinder becoming a torus above). It is this same feature that leads mathematicians to identify coffee cups with doughnuts: they are topologically identical, each having only one hole or handle. So the predictions of general relativity in a coffee cup universe are identical to those in a doughnut universe so long as the metric field is transformed in the same way as any matter fields by the diffeomorphism that brings about this shape shifting. Likewise, the predictions will be identical regardless of whether the world is shaped like a ball or a bowl.

The Einstein Shift

This idea in general relativity that if one replaces the spacetime manifold with a topologically equivalent (i.e. homeomorphic) manifold, then the physics 'stays the same' is at the root of the hole argument – originally, it was the ability to use any coordinate system to describe some physical situation (i.e. the passive understanding of the symmetry). Since homeomorphic manifolds do not differ in their topological properties, and these are what matters, the observable content of the theory is unaltered by the action of a diffeomorphism. This feature has forced many to give up on the idea of a spacetime 'sitting under' physical events, and the hole argument plays a large part in this. So let's present a simplified version.

We can run something like the Galileo ship argument in this case too. Now imagine again that a rather more powerful being, such as the Flying Spaghetti Monster, wants to fool you. First, you make a bunch of measurements using all the machinery we have, to determine a model of the universe, with some spreading of the fields onto the manifold. The Spaghetti Monster then puts you to sleep, does some reshuffling of the points of the manifold (or smearing of the fields over those points) and then wakes you up. If the points are real then the Spaghetti Monster has generated a physically distinct situation. But you won't be able to tell what has happened, since all observables (those physical quantities that satisfy the laws of general relativity) will be the same, since all the monster did was apply an operation relative to which the laws are insensitive – shifting matter *and* metric by the same transformation – in which case so are the observables. The formal foundation of this lies in the fact that we are now moving structure that was fixed in the context of the Leibniz shift argument (the metrical structure) *together* with the matter fields. This means that what look like very significant changes, that warp and bend things out of shape, are not detectable, much as doubling the rate of *everything* (includings one's means for checking on the rates of change of processes: clocks, pulses, orbits, etc.) would leave things looking just as they did before.

A little more technically now: we like to think of spacetime in relativistic theories as a four-dimensional block, but if we want to look at dynamical features, it's useful to carve this block up into three-dimensional slices. We can do this in general relativity, so long as we realize that our slicing has no real physical significance and many such slicings (which would lead to the same four-dimensional block) are possible – this is known as 'foliation invariance.' Suppose that we know the matter distribution and the geometry with absolute precision up to and including some slice $\mathcal{S}$. The freedom to perform a diffeomorphism means that even though we have specified everything up to and on our slice (roughly representing our 'Now'), infinitely many possible developments of the fields off that slice are possible.

As we will see in §5.3, having multiple possible futures from some initial conditions amounts to indeterminism, and so it appears as though general relativity itself is indeterministic. For example, by running the initial state through the equations of general relativity (the laws), we might generate the world represented by the model $\langle \mathcal{M}, g, T \rangle$ or we might generate the world $\langle \mathcal{M}, \phi^*g, \phi^*T \rangle$ (where ϕ is a diffeomorphism, of which we have infinitely many to choose from, and ϕ^* an operation that drags fields, such as g and T, around over the points of the manifold).

The catch is that the various apparently possible futures only differ with respect to which points are sitting under which field-values. So we want to know whether $g = x$ or $\phi^*g = y$ sits at some specific point p. As mentioned, no observable facts are affected by this indeterminism. So why on Earth should we be concerned? In the philosopher's version of the hole argument, due to John Earman and John Norton [8], spacetime substantivalists ought to be concerned, since they believe in the reality of the spacetime manifold and its points. If we follow this view then there should be a fact of the matter as to which point the Spaghetti Monster shifted some field value to, even though it is opaque to experiments. This is, of course, extremely close to the Leibniz shift argument, only with the more general diffeomorphism group taking the place of the Galilean group. Just as Leibniz thought that the proliferation of possibilities that realism about space and time generated in Newton's world amounted to a demonstration of its absurdity, so Earman and Norton argue that substantivalism must be rejected, for reasons of physics: general relativity is not indeterministic in any sense that matters, and an interpretation that says it is should be rejected. In their own words:

> Determinism may fail, but if it fails, it should fail for a reason of physics, not because of commitment to substantival properties which can be eradicated without affecting the empirical consequences of the theory. ([8], p. 524)

The alternative is associated with Leibnizian relationalism: view all of the diffeomorphic futures as representing one and the same physical possibility – they call this "Leibniz equivalence." The Spaghetti Monster was fooled into thinking she was doing some non-trivial operation by the mathematical machinery we use to talk about the world according to general relativity! In reality, so this Leibniz equivalence option goes, the mathematics of such transformations is a piece of representation that, while helpful in many ways, does not map onto the world: the world is best represented by the (intrinsic) structure that is invariant under such transformations – this equivalence class of diffeomorphic metrics is called the *geometry* by physicists, by contrast with the metric.

Earman and Norton wrote their paper in 1987, and it sparked an explosion of papers and alternative views, some seeking to defend substantivalism, some accepting the relationalist thrust. We will sample a few of these here, but I leave it to you to decide which makes best sense.

Getting out of the Hole

Firstly, we need to say something about the extent to which endorsing Leibniz equivalence (i.e. the idea that general relativity is about diffeomorphism equivalence classes of metrics) is in fact relationalist. Historically, of course, accepting the idea that a bunch of symmetric possibilities represent the same state is associated with relationalism, as we saw in the Leibniz–Newton debate. But one immediate problem with this view is that general relativity in fact allows for so-called 'vacuum solutions' to the field equations, meaning that there is just pure gravity in such worlds (nothing we would ordinarily call matter). So we must ask ourselves, how can there be a relationalist interpretation of entirely empty space? Indeed, how do we contend with the fact that general relativity is a theory *of* spacetime, and so is presumably committed to its existence?

The catch here is that, as we suggested earlier, spacetime in general relativity (the metric field) is a rather different beast to spacetime in all prior theories, and this has to do with the multitasking feature (spacetime doubles as the gravitational field), and as the quantum gravity theorist Carlo Rovelli likes to put it, with a strong enough gravitational wave, you could smash a rock to pieces. Since the metric field is everywhere defined, if it is as substantial as it seems, then the relationalist has something defined all over and needn't worry about empty space. But, the substantial entity is spacetime, so why is this the property of the relationist rather than the substantivalist? This confusion (among others) has led the philosopher Robert Rynasiewicz [41] to dismiss the substantivalist versus relationalist debate as "outmoded" in the context of general relativity since the categories used in the original formulation no longer make sense. This makes sense: if both sides are claiming that the self-same object is real then what are they fighting about?

However, it is possible to restructure the debate in the light of the new developments so that we can still have a meaningful debate about how general relativity maps to the world. Carl Hoefer [24] has argued that a version of substantivalism fit for general relativity can be constructed and, given that relationalism is just the denial of substantivalism, so can an account of relationalism in general relativity: it is just the rejection of the idea that spacetime is part of the theory's fundamental ontology – though, he argues, the latter is not necessarily as well supported as the former and it might well be that the debate is more or less *settled* in favor of substantivalism. The idea is simply that the metric field plays the same role in this new context that the Newtonian container played in the classical debate: the difference is that this new spacetime is not inert, but influences matter and is backreacted on by that matter.[18] This change does not affect the fact that the metric provides us with our basic spatiotemporal facts. Either these spatiotemporal facts are grounded in a real substantival spacetime as modeled by the metric field, or they are grounded in something else.

However, we must not forget that this can be no direct mapping from metric field to spacetime: a lesson of the hole argument is that the metric at a point is itself not physical, since we can smear it arbitrarily over the manifold without changing the physical possibility described. What is physical is, instead, the equivalence class of such smeared metrics (the geometry). But this move as we saw was associated with relationalism. A reaction from substantivalists has been that since such an equivalence class would simply encode the intrinsic physical structure of spacetime, there is no reason why they too shouldn't help themselves to it – this is known as 'sophisticated substantivalism'! But, if both relationalists and substantivalists are invoking the same structure then where does the difference lie? Hasn't the distinction between relationalism and substantivalism simply collapsed? Not quite. On the surface, the substantivalist has more of a case for claiming that the structure corresponds to a truly existing spacetime than the relationalist has for saying that it is in some sense generated by relations. But it is not clear cut. New work, originating in research on quantum gravity, argues that the observables (invariants) of general relativity are necessarily relational (taking the form of correlations between field values).

Earlier responses to the hole argument attempted to prop up substantivalism by consideration of modal metaphysics (having to do with possibilities and possible worlds), some of it quite arcane. To see how these work, bear in mind that the distinct models (or worlds, if you prefer) that are generated by diffeomorphisms differ only with respect to which (invisible) manifold point plays host to which (visible) feature. In one world the point p might be host to the location of maximum curvature, while in a diffeomorphic version that role is played by the point q. This kind of non-qualitative difference (amounting to an invisible role-swapping) is known as a 'haecceitistic' difference: the same individuals (points and fields) are present in both worlds, and exactly the same observable relations are realized, but by different individuals in each case. One approach, due to Tim Maudlin [32], to saving substantivalism suggested that points might wear their metric field values as essential properties, so that a world in which they don't have those selfsame properties is simply not a genuine physical possibility – this only works in situations where the diffeomorphisms do not preserve the points' metric properties (i.e. where they are not 'isometries'), so that a world with symmetries is excluded. There is a simple and cogent objection to this view, which is that, while we might agree that metrical properties of *some kind* are essential to spacetime points, to rigidly attach just those metrical properties a point *happens* to have as a matter of fact (in our world) seems to be too strong. For example, we can't talk about fairly innocuous counterfactuals that involve the point having different properties, such as 'if I hadn't made a cup of tea five minutes ago, the curvature around my desk would have been different' – this seems to commit us to the necessity of my teacup being on the desk since the points within the desk and cup would take on different metrical properties!

Maudlin's response is that we can help ourselves to modal talk of this kind, but without invoking the *same* points clothed in different properties, by using a tool associated with modal logic, known as counterpart theory (due to David Lewis). This says that the statement 'if I hadn't made a cup of tea five minutes ago, the curvature around my desk would have been different' is true because there is a counterpart desk, cup, and point with these properties. Jeremy Butterfield [3] simply bypasses the metrical essentialist component and uses counterpart alone to motivate a defence of substantivalism: there is only one world with my desk, cup, and the points they occupy. Again, we can consider modal facts about them, but this need not involve those objects being the same in the possible worlds considered: the points in a spacetime and a diffeomorphic version are not the same since the counterpart relation is different from the identity relation. Indeed, choosing a good counterpart relation would involve choosing the closest match for some point in the other scenario, and that would be the one to which the fields were dragged by the diffeomorphism.

Other responses work by similarly denying that there is 'transworld identity' linking the points in the different solutions (in a non-qualitative way), but without the additional modal gymnastics. If we simply deny that points have some kind of primitive identity that transcends their qualitative properties, then we end up achieving Leibniz equivalence through the back door. The points of the manifold aren't transported from world to world, forming an absolute background: if we want to know what points are the same across worlds we look at their qualitative properties. This is the basis for the sophisticated substantivalism mentioned above. Simply put: haecceitism need not be viewed as part and parcel of substantivalism as Earman and Norton had suggested. The problem is, however, that this leaves us very little room to distinguish relationalism and substantivalism, as before. It is possible that a view that simply merges these positions might be more favorable.

4.5 Further Readings

There are a great many books on both the physics and the philosophy of space, time, and spacetime. Many of the latter can often depart from the physics, and lie more within metaphysics than philosophy of physics.

Fun

- Edwin Taylor and John Wheeler (1992) *Spacetime Physics: Introduction to Special Relativity* (2nd edn). W. H. Freeman and Company.

  - This remains one of the best textbooks for beginners to gain some actual computational feeling for special relativity in a light-hearted way – it helps that John Wheeler was one of the great physicists.

- Nick Huggett, ed. (1999) *Space from Zeno to Einstein: Classic Readings with a Contemporary Commentary*. MIT Press.

  - Very useful collection of many of the ancestral voices of contemporary philosophy of spacetime physics, including some of the original papers corresponding to topics discussed in this (and the next) chapter (by Leibniz, Newton, and Kant, for example).

Serious

- John Earman (1989) *World Enough and Space-Time Absolute vs. Relational Theories of Space and Time*. MIT Press.

  - The classic treatment of the debate between substantivalists and relationalists. Exceptionally clear, full of good sense, and still relevant.

Connoisseurs

- Jeremy Butterfield, Mark Hogarth, and Gordon Belot, eds. (1996) *Spacetime*. Dartmouth.

  - It will cost you an arm and a leg to buy, but this is truly a dream collection of pivotal papers on themes discussed in this (and the next) chapter. One can gain a very good feel for the field of philosophy of spacetime physics from this one text.

5 Further Adventures in Space and Time

This chapter steps away from philosophical issues stemming from symmetries but stays firmly focused on space and time. The three topics we cover tend to be of a more epistemological flavour than the previous chapter's ontological problems: we begin with a look at the idea of the 'true geometry of the world' and consider whether we could ever discover such a thing. We then consider a similar problem involved in the idea of measuring time and finding a 'true time.' Finally, by way of also limbering up to the next chapter on statistical physics, we consider the status of determinism in physics.

5.1 Can We Know the World's Geometry?

We tend to think of the world as having some definite geometry, and we might also tend to think that this geometry is one of those things that scientific work can help us discover. For example, depending on what the geometry of space is, the internal angles of a triangle will be more, less, or equal to 180 degrees. If only we could make a big enough triangle, we could test this. (In fact, Carl Friedrich Gauss is reported to have performed such an experiment in the 1820s by measuring the angles of light beamed between three peaks in Hanover – whether this experiment was *really* supposed to constitute a test of the deviation of the world's geometry from Euclidean geometry is a matter of debate among historians of mathematics.) Likewise, for other plane figures such as squares and circles, with the measured properties altered accordingly. It is just a matter of measurement. Or is it?

Poincaré's Parable of the Surveyors

Henri Poincaré famously invoked a kind of 'discworld' (long before Terry Pratchett, and also before general relativity came along with its curved spacetime) that any number of (mutually inconsistent) world geometries could be made consistent with our observations of the world, including our direct sensory experience.[1] A team of flatland surveyors is confined to a closed Euclidean disc (i.e. with an edge at radius R), armed with rigid rods and light rays to make their measurements. He then adds a temperature dependent rod length and light refraction in this world and makes the temperature fall off as one strays from the disc's center, with distance ρ. (Note that he actually encloses them in a large sphere, but his discussion suggests taking a cross-section of the disc through the center, with a radius R in which the distance of one of the inhabitants ρ is measured from the center. The temperature is then proportional to $R^2 - \rho^2$.) All objects in this world dilate and contract by the same amount ($R^2 - \rho^2$) and the thermal equilibrating effect happens at an instant. As one probes further out from the origin the rods contract more and more, becoming smaller and smaller (as well as colder, though the flatlanders wouldn't be able to measure this since their thermometers suffer the same distortions). The surveyors know nothing of this distorting force, naturally assuming their bodies and instruments were rigid on account of feeling and observing no such effects. A similar force afflicts light rays, which have an index of refraction *inversely* proportional to $R^2 - \rho^2$. In modern discussions, we speak more generally of 'universal forces' rather than temperature: all we need is to postulate a force that dilates objects uniformly in the same way so that it goes completely unnoticed.

Of course, in a flat Euclidean world, the ratio of the circumference to the radius is simply 2π. However, with the distorting forces of the temperature or whatever universal field one postulates generating the same behavior, our surveyors, in figuring out the intrinsic geometry of their world (using whatever tools we might use to do the same: string, rulers, lasers, etc.), will find values *greater than* 2π characteristic of a hyperbolic (Lobachevskian) geometry (i.e. one with negative curvature). Likewise, measured triangles will have internal angles adding up to less than 180 degrees. The effect will be more dramatic as one measures larger and larger radii, circumferences, and triangles. Moreover, since their measuring instruments would shrink as they approached the boundary, they would never reach it (see fig. 5.1). From their results they would (wrongly, by construction: we know it is a finite Euclidean disk) infer that they live in an infinite non-Euclidean world – if distance is defined in terms of what is measured with rulers and the like, then the space is infinite in extent! They have, of course, wrongly it transpires, a (perfectly rational) 'rigid body hypothesis,' which ensures that merely moving about in space will not distort shapes and sizes.

But now suppose that maverick physicist Albert Fleinstein (the flatland counterpart of Einstein) points out that all of the surveyors' results are compatible with the presence of precisely the forces introduced above in a flat, closed Euclidean world. So we have two theories:

T1   The world is infinite and non-Euclidean (hyperbolic).
T2   The world is finite and Euclidean, though with universal forces.

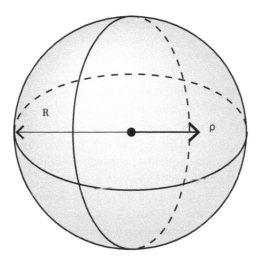

Fig. 5.1 Poincaré's surveyors are enclosed in a sphere of radius *R*. As they move a distance ρ from the center, objects that were, e.g. 1 meter long at the center will be just $(R^2 - \rho^2)/R^2$ meters long – the boundary is unreachable since then ρ = *R*, so that $(R^2 - \rho^2) = 0$. Hence, the space would be deemed infinite by beings confined within its borders.

The problem Poincaré poses is, how can Fleinstein's fellow discfolk decide from within their world which theory is correct? There is no 'stepping out' of their two-dimensional standpoint, to our God's eye view, to check. According to Poincaré the question is simply undecidable by experience or reason, and must simply be stipulated as a matter of *convention*. The problem is that any experience or experiment that makes one theory true will also make the other theory true. We have what philosophers of science call 'underdetermination of theory by evidence': the evidence can't decide the matter. But it isn't a case of simply not having gathered enough data: no possible data, consistent with the construction of the simplified world, can settle the controversy since it will be derivable from either theory.

The conventional choice could be made for various reasons (simplicity, coherence with their other theories, closeness to 'the experienced world,' etc.). But there is no absolute correctness to either choice since there is no absolute criterion on which to base it. A convention is just another name for an 'implicit definition' (an arbitrary choice of the language employed). For Poincaré facts of geometry are conventions in this sense: free and bounded only by the avoidance of contradiction. Poincaré himself believed that the flatlanders would be best served by choosing the 'Euclidean space + forces' option, invoking its superior simplicity and closeness to our everyday intuitions about space.[2]

> Our choice is not .. imposed by experience. It is simply guided by experience. But it remains free; we choose this geometry rather than that geometry, not because it is more *true*, but because it is more *convenient*. ([36], p. 145)

Of course, we can easily quibble with this, pointing out that these curious forces are rather messy, while using the geometry of space (to 'embody' or 'geometrize' the force) is somehow more elegant and unified. This is not really relevant, however: the key point is that one can apparently gerrymander a finite flat space picture with some curious distorting forces to capture all of the empirical facts that a geometrical picture with an effectively infinite negatively curved space might generate. In which case, we cannot be said to know the world's *true* geometry. In each case, we define the terms of the theory in such a way that the laws or axioms (e.g. of geometry) come out true.[3]

A Topological Parable

Hans Reichenbach ([39], pp. 63–66) considers a similar example involving beings that live on the surface of a sphere. One can, he argues, once again generate a parallel story, concerning *topological* features of a space, by this time redefining aspects concerning the *reidentification* of objects in the space. One might think it is perfectly easy to tell what the shape of space is in this case by simply planting a flag in the ground and traveling far enough to return to the starting point and finding your flag. Reichenbach points out, however, that this depends on a convention about objects 'being the same.' One might have a situation in which one is in a flat space but as one moves out certain features (such as the flag you planted) are mysteriously duplicated so that it only looks as though you are back where you started. There would need to be some principle of 'pre-established harmony' in the overall space, where the other stretches of space 'know' that the flag was placed a certain distance away so that it could be duplicated accordingly. This example is harder to uphold if we imagine stretching a very long rope around the space so that one could tie a knot in it. The sphere's surface would provide an obstruction when pulled tight, which should be missing in the flat infinite space with duplication. However, this is too quick: presumably one must have tethered the rope to something (a tree say), which essentially is no different from leaving a flag. All one would see, as one completes the journey, is an end of a rope tied to a tree: there is no certainty that it is the other end of *your* rope! It might simply lead off in a straight line again, onto the next clone of your world.[4]

It is clear that this kind of thinking can be ramped up a dimension so that we imagine the same scenario occurring in a universe just like ours – of course, Poincaré was using his case as a possible analogue for our world.

We can ask: is the world open or closed (by analogy with the surface of a sphere)? If it is closed one could imagine setting off from the Earth keeping a straight line course and eventually returning to Earth. We face the same reidentification dilemma: is this the same Earth you left or an identical one some great distance from the 'real' Earth? If you think that it is a different Earth then you have to accept all of the strange coincidences (your cup was left in the same place on the desk in your office as the cup here on this imposter desk in this twin Earth's version of your office). You can try to 'catch' the twin Earth (or some other copy) out by leaving a special message locked in a safe where only you have the combination. But you travel once again and come to find the safe opens with your code and the same message is in there: it is an assumption (though perfectly reasonable) that this is the *same* safe and message; the pre-established harmony story would have the same observational consequences. The more acceptable alternative is, of course, that you live in a universe with '*Asteroids*-geometry' (a toroidal structure, so that going far enough in one direction brings you back to your starting point). In other words, the kind of *topological* structure (whether a space is open or closed for example) depends to a certain extent on our preference for a good causal story, with no spooky influences, such as the curious duplicating of one world in another location.

Reichenbach ([39], §17) has also extended this to other scientific facts at the basis of our theories, such as the uniformity of time (relating to the metric of time), which is based on the idea that we *stipulate* (by a 'coordinative definition') that, e.g. a pendulum's swings cover equal periods of time – this is based on the fact that we can't compare successive durations, there is simply no way to test such a thing:

> We cannot carry back the later time interval and place it next to the earlier one. It is possible to make empirical statements about clocks, but such statements would concern something else. Two clocks stand next to each other, and we observe that the beginning as well as the end of their periods coincide. Further observation may show that the ends of their periods always coincide. This experience teaches us that two clocks standing next to each other and having equal periods *once* will *always* have equal periods. But this is all. Whether both clocks require more time for later periods cannot be determined. (ibid., p. 116)

This leads to a curiosity in physics (one that we return to in §5.2): the laws of physics themselves suggest that such periods will be equal, but those very laws were the result of experience with clocks "calibrated according to the principle of the equality of their periods" (ibid.). This is a very tight logical circle! To break out of it, says Reichenbach, requires an acceptance of conventional elements in the measure of time. Again, according to Reichenbach, any such definitions are chosen for the way they simplify description. This does not point to their truth, but identifying such conventional elements (given that they are contributions from the mind) is an essential part of separating out subjective from objective structure in our descriptions of physical reality.

Realism versus Conventionalism

This brings crashing home the point that conventionalism has a tendency to align with anti-realism about whatever is subject to the conventionalist stance. To say that something is conventional is to remove it from the objective world. However, what it also reveals, if we accept it, is that scientific theory-choice goes beyond empirical evidence in such cases.[5]

More recent work within the area of philosophy of cosmology has tended to focus on the epistemological opacity of various features of spacetimes (in general relativistic universes), which implies that we can never be fully sure of the structure of the universe: there are multiple consistent (but unobservable) developments of the observable part. The issue arises from the existence of *causal horizons* beyond which we can't have knowledge. This is rather different to the kind of conventionalism mentioned above: there might be a fact of the matter, but it is simply our empirical limitations (i.e. restrictions on what we are able to experience, observe, and measure) that prevent us from finding them out. Unless we consider not being able to probe higher dimensions as an 'empirical limitation' (which seems wrongheaded in any case), the cases discussed by Poincaré and Reichenbach transcend empirical matters.

But what picture of the world are we left with then? One with an 'indeterminate' geometry and topology? Or a world with a definite geometry and topology, but that will lie forever from our view? Why should we care in any case? There are many other conventions (driving on the left side of the road in Australia) that we do not fret about: is it 'really true'! The difference is that we can witness other conventions, and imagine changing them and doing otherwise with visible effects. Changing the convention would make a difference to the world. But that might cause us to be even less impressed by these geometrical examples: if the choice has no observational impact whatsoever then is it a difference worth worrying about? Perhaps a better example, used by Poincaré, is that of different coordinates (Cartesian versus polar) or the choice of units in the making of measurements. We often have to switch from pounds to kilograms, or stone, because of the different conventions for weight measurement in different countries. We don't think that one is 'more correct' and yet these do not make a difference to the measured quantity. We also have the element of convenience of a unit relative to purpose, or providing a better fit with other units, and so on. Likewise there are all sorts of ways of measuring temperature (Fahrenheit, Celsius, etc.), but though there might be a disagreement about the numerical value given, there will be no disagreement about the qualitative aspects: the chicken in the oven will cook in the same amount of time regardless of whether we have the setting at 350 degrees Fahrenheit or 176.6 degrees Celsius. We simply fix some set of units to know what we're talking about and to be able to specify what to do in a recipe. They give us, as beings that interact with the world and each other, a *grip* on temperature. Nobody imbues the units with any physical significance beyond their convenience. Again, we seem to be back to anti-realism about conventional elements.

Clark Glymour [18] has argued that even when there are conventionalist choices to be made we need not be forced into anti-realism. There are often reasons to say that while we cannot decide the matter, there is nonetheless a fact of the matter – these situations occur more in the kinds of cases where we are empirically constrained from finding certain things out. He also suggests that the deadlock can be broken with solid methodological considerations beyond empirical factors [19]: empirical equivalence does not mean equivalence in all scientifically relevant respects. General relativity (Einstein's theory of gravity) also causes some problems for the "free to choose" idea where geometry is concerned (topology is a different matter) since there the field equations involve a dynamical interplay between the geometry and the matter distribution such that they are bound together with the latter seemingly *uniquely* selecting the former – in §4.4 we saw that this isn't quite so straightforward as is often supposed.

Another escape from anti-realism is to argue that, as with units, we know that there aren't really two separate 'theories of the world' being offered: one and the same physical content is represented by both systems. This suggests treating a theory as a kind of equivalence class of its observationally identical presentations: theories are systems that tell us what we will observe and explain what we have observed. This is associated with the 'positivist' school in philosophy of science according to which what is not observable (such as the difference between the conventionalist scenarios presented by Poincaré) is strictly speaking *meaningless*. If we don't want to go down this path then we have to say something about the ontological nature of whatever it is that the two theories are redundantly representing, about which positivism remains silent. One realist option, due to Adolf Grünbaum [21], argues that what is shown by the geometric underdetermination cases is simply that space is 'metrically amorphous' (lacking in intrinsic metrical structure) so that Poincaré is seen to be right that geometry is conventional, but this needn't lead us into anti-realism itself.

'Structuralist' positions will point out that the structure revealed by the equivalence class (i.e. whatever is common to both descriptions) exhausts what we can know (epistemic structuralism) or, in more extreme versions of structuralism, exhausts what there is (ontic structuralism). Another option that fits well with such cases in which there doesn't seem to be a fact of the matter about which is correct is 'constructive empiricism' (due to Bas van Fraassen). This is realist about observables (on which the two theories match) but agnostic about the unobservables (on which the two theories do not match). However, to be pushed into such extreme (and global: applying to all theories) positions by a cluster of theories might be going too far. One can potentially rescue realism from some conventionalist dilemmas so long as there is a 'dictionary' linking the respective theoretical structures, as well as the matching of the structure of observables. This is hard to deny, but there might be some problem cases that slip through the net, in which the theoretical structures are simply too heterogeneous to be mapped onto one another in the required way.

More recent work, especially that occurring in string theory, has raised the spectre of conventional aspects in physics once again. Transformations known as *dualities* between (what appear to be) physically different string theories lead to the same observable content. One simple yet striking example is 'T-duality.' Here a string theory defined on a space with a large radius r is indistinguishable (using strings and the laws they obey) from a string theory defined on a space with a small radius, $1/r$ – the details needn't concern us here. One possible response is that the radius is conventional just as in the geometrical structure of discworld. However, there are other options here, matching those above: we might remain agnostic about the issue, though perhaps still accepting that *one* of the radii correctly describes the space. Or we might take the theories as simply different ways of talking about the same physical possibility?[6] If we follow this latter route then it seems hard to retain the naive picture of the world as strings living in spacetime.

5.2 Measuring Time

If you wear a wristwatch, then as its battery approaches expiry you will notice that it 'slows down.' If you're like me, then this slowing down is gauged relative to your laptop, which has its time set (I'm told) by an atomic clock: we assume that this atomic clock is more reliable and so if there is any drift between watch-time and the laptop-time, we can usually safely assume the problem is the former.

Without this kind of comparison (and assumption), how do we judge whether a clock is slowing down or speeding up? How do we tell whether a pair of time intervals are the same or different, which is what is required? After all, we only experience the world as it unfolds. We can't measure Newton's (invisible) absolute time, and even using 'sensible measures' (clocks of various kinds) we can't archive the intervals that have passed (tick-tocked) in order to compare them. Unlike spatial distances, in which we can place objects side by side, intervals are one-off entities. To return to the watch versus laptop example again, how do we know that it wasn't a case of the laptop clock speeding up because of some fault? The two scenarios would be identical: the relative separation between the times shown doesn't care what scenario causes it, the slowing down of one or the speeding up of the other.

A Convenient Time

Poincaré identified this as a key problem with time measurement, one that is both practical and philosophical – it is in many ways the temporal analogue of his geometrical conventionalism.[7] The problem is that while we can with confidence state when events are before, after, or simultaneous with one another (topological ordering), it is not so easy to state when two intervals of time are identical (the metric properties: the *how much*): we can't just 'sense' such a thing. He didn't have the luxury of a laptop set by an atomic clock, but he uses a similar example:

> Of two watches we have no right to say that one goes true, the other wrong: we can only say that it is advantageous to conform to the indications of the first. ([37], p. 228)

We might use a pendulum, for example, and assume that its beats are all of equal duration, but we know that there are all sorts of irregularities caused by temperature, air pressure, and so on. Correcting from these (and subtracting them somehow) would still leave the equality approximate, since there are electromagnetic influences and even tiny gravitational perturbances from other astronomical objects beyond the Earth. The pendulum clock is so prone to disturbance that the Earth's rotation itself was used as a watch instead, so that each full rotation is a tick assumed to have the same duration.

But this new watch has its own problems. There is a slowing down of the Earth's rotation due to the tides (and other influences), which results in a measured speeding up of the Moon's (and other bodies') motion relative to the Earth's 'ticks' (when combined with Newton's laws of motion). The Earth's slowing down, however, is measured from the Moon's apparent speeding up! The observed acceleration of the Moon would be in conflict with Newton's law and conservation of energy if the Earth's rotation were taken to be uniform, so an appropriate correction is made, attributing a deceleration to the Earth.

But, as Poincaré points out, this puts the weight on Newton's laws, which are also approximate, as empirical facts. Moreover, with this definition of time based on Newton's laws, we could pick any periodic phenomenon as our watch, and so long as we made the appropriate corrections, so that any observed feature remains consistent with Newton's laws and the conservation of energy, we have much the same principles at work. Some such watch might, however, result in very complex corrections and a messy statement of Newton's laws. This is the key for Poincaré; as with his disc-world, he argues we tend to adopt the more *convenient* standard of time measurement, rather than the 'most true':

> Time should be so defined that the equations of mechanics may be as simple as possible. ([37], pp. 227–228)

The similarity to the discworld case should now be clear: we have options for either sticking to one set of laws (or, in Poincaré's terms, one "enuncia-tion" of the laws) or choosing some other more complex statement. We can say that the Earth is perfectly uniform (modulo corrections for tidal fric-tion and other influences, knocking it off its true course) and makes a fine *t* for Newton's equally fine equations, which leads to accelerated motions in systems referred to it, or find a more suitable (more uniform, without corrections) periodic phenomenon.

There is something strangely circular about all this: we can choose to take any one of the planetary objects as a clock, and imposing Newton's laws (and solid principles of physics, such as the conserva-tion of energy), by assuming those laws and making observations, we make whatever corrections to our clock as are needed to get the whole system consistent. A choice of object is made purely to achieve the great-est simplicity of the form in which the laws are expressed. The time *t*, then, that features in Newton's equations, is *defined* by those very laws (together with observations that are supposed to be used to confirm the laws)! This has much in common with pulling yourself up by your own bootstraps.[8]

New Standards

A more robust watch is the atomic clock, which is far less susceptible to external perturbations – theoretical calculations show that its various beats are uniform (identical in duration) that it will lose only a second in tens of millions of years.[9] But, a second is still an irregularity, and the same procedure must be adopted in our scientific practice: assume some laws of physics (not necessarily Newton's), add our observations, and then figure out how the system realizing the *t* in the equations will need to be adjusted to make the observations and the theory consistent.

There are two opposing interpretations of this procedure (i.e. in terms of *what* clocks are measuring and how *t* maps to the world): realism and anti-realism (conventionalism) about time. We have already seen the latter: Poincaré's claim that there is just no fact of the matter about what the 'true time' is, only choices that result in simpler and more complex formula-tions of the laws. But much of the terminology of 'corrections' to the time variable at least suggest an underlying true time that our advancing, ever more *precise* choices of clock are approximating: better clocks in this sense are those that map more faithfully onto Newton's true time.[10] John Lucas explicitly adopts this viewpoint:

> The fact that we have a rational theory of clocks vindicates Newton's doctrine of absolute time. If we really regarded time simply as the measure of process, we should have no warrant for regarding some processes as regular and others as irregular. ([30], p. 91)

However, precision (to a greater or lesser degree) does not necessarily mean more accurate (in terms of mapping onto some quantity in the world: absolute duration). Sklar considers the relation as a *causal* one, rather than mapping (though only to dispel such a notion):

> Of course, deviation of any clock from its ideal rate is something to be explained by causal interaction in the material world. But there is no "causal" explanation as to why clocks in general record time intervals more or less accurately. What we mean by time intervals is just this numerical abstraction and idealization from the uniformity more or less of relative rates of clocks of various kinds of construction. It is, of course, still an important observation of Newton's that only when we date events by the ideal time metric will our dynamical laws of nature take on their familiar simple form. But that does not seem to call for absolute time as a "cause" either. ([47], p. 74)

Precision can also refer to an ability to control the various errors and perturbations, at least with theoretical knowledge on how to remove them from calculations. Moreover, we have seen that the conventionalist is perfectly capable of biting the bullet and accepting that we really do not have any such (absolute) warrant for distinguishing regular from irregular.

There have been several important advances in time measurement since the turn of the twentieth century. Firstly, there was 'ephemeris time,' which essentially followed the idea that since the solar system could be viewed as a kind of clockwork machine, it should be used to define time – 'ephemeris' refers to the catalogue (an 'almanac') of positions of some astronomical object over time. The 'hands' of this clock are the positions of the Moon and planets (relative to the 'fixed stars'), as determined by Newton's laws. Ephemeris time is then just the rate at which these 'tick.' The unit in this case was the sidereal year (as of 1900: to avoid inevitable fluctuations), or whatever time it took that year for the Earth to perform a complete orbit around the Sun.

The next step was the creation of atomic time. A clock, of course, is simply something that oscillates (preferably in a uniform fashion, modulo the problems raised above), along with a register of the number of cycles that have occurred. The specific oscillator provides the 'frequency standard' (e.g. a pendulum, a pulse, the Earth's rotation, the solar orbit, etc.). Atomic time is based on the recognition that atoms vibrate at specific frequencies, and so it involves an atomic frequency standard. This nicely fits the natural criterion of a standard, that it be 'universal' (freely recreatable wherever and whenever one wishes). Atoms of the same kind are identical, unlike pendulums. The atomic second is, however, defined in terms of the second of Ephemeris time: a second of solar time is correlated with 9192631830 ± 10 cycles of a caesium atom.

The study of standards is a fascinating one, and hasn't received nearly enough attention from philosophers of physics.[11] However, I raise it here to simply point out that the same philosophical issues are raised regardless of the standard we use. There is the same question of *what* is being measured: a 'real time' or simply physical processes that are linked to the clocks via correlations. However, a future advance (still a 'work in progress') will attempt to base a set of standards (for time, space, and mass) purely on the fundamental (universal) constants of nature: Planck's constant $\hbar$, the constant of gravitation G, and the speed of light, c. I leave it as an interesting exercise for you to figure out what difference (if any) this change in standards would make.

5.3 Determinism and Indeterminism in Physics

The most famous characterization of determinism (indeed, amounting to the very definition of the claim for most, and quoted whenever the word 'determinism' is mentioned) is due to Pierre Simon de Laplace:

> We ought to regard the present state of the universe as the effect of its antecedent state and as the cause of the state that is to follow. An intelligence knowing all the forces acting in nature at a given instant, as well as the momentary positions of all things in the universe, would be able to comprehend in one single formula the motions of the largest bodies as well as the lightest atoms in the world [so that] to it nothing would be uncertain, the future as well as the past would be present to its eyes. (In [33], pp. 281–282)

Hence, the present state of the world is understood to have been 'brought into existence' by a unique prior state together with the laws of nature (Laplace's 'forces'). We have in here, then, laws of nature and the relation of cause and effect. We also have the notion of *predictability* (in principle), as based on these other elements. It is not surprising that this vision of a deterministic universe was couched in the framework of the clockwork-conceived Newtonian solar system, with the planets, Sun, and Moon linked by gravitation. For example, given the initial positions and velocities of all particles together with Newton's second law $\mathbf{F} = m\mathbf{a}$, then so long as we know all forces $\mathbf{F}$, given some mass, we will be like Laplace's 'intelligence' (or 'Demon') in terms of computing motion. Because of the interlocking nature of the forces between all of the objects in the system, one has the potential to know its state at any instant one could care to choose.

In more modern terms, then, determinism simply means that for some initial condition, given the laws (and any boundary conditions), there is *one and only one possible outcome* (relative to those laws). We can represent this diagrammatically as follows:

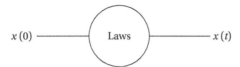

Note that here $x(t)$ might be either to the past or future of $x(0)$ (the initial state): hence, the idea is that the laws and an initial state will determine a unique *history*. This can be further transformed into a statement about *replicating* initial conditions, since it follows that whenever some state is reproduced, it will duplicate the behavior of the original. like causes will have like effects, in other words (like replaying a videotape). The phenomenon of *chaos* often fools people into thinking that it implies a failure of determinism since we lose the ability to predict future states (over certain timescales). However, determinism (in the above sense) is preserved, only the ability to replicate initial conditions is lost, and the laws are such that even small errors in this replication will be pushed into large divergences in later states by the laws. Indeed, in general, if we don't have a perfect grip on the initial conditions, we will pick up some uncertainty in how the system will evolve.

Indeterminism, by contrast, means that for some initial condition, given the laws, there is more than one possible outcome (though only an individual outcome might be found).[12] In this case, we can see that the 'like causes will have like effects' principle is violated: we can duplicate the law and the initial conditions, and yet get different behavior. This can be represented by a branching structure as follows:

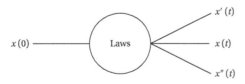

We saw in the previous chapter (in the context of the hole argument) how the laws of general relativity have some freedom so that the same initial state could lead to what looked like distinct future states, but those states formed an equivalence class under the theory's central symmetry, which would collapse the branching structure. However, this should lead us to suspect that deciding whether a theory is deterministic or not isn't quite as simple as one might believe: in that case it was a matter of interpretation whether one collapsed the branching possibilities (that is, whether one viewed them as representing physically distinct possibilities).

Uncertainties and Probabilities

Of course, this branching situation might also reflect uncertainty in terms of our knowledge of the outcome rather than any uncertainty on the part of Nature. A toss of a die is an obvious case in which the uncertainty is in our heads, and if only we knew "all the forces acting in nature at a given instant" on the die, we would be able to compute its outcome. There, of course, we have a branching into six possible outcomes. Of course, though we cannot predict with certainty which of the six outcomes will be realized following a throw, we can in this case have some say over the *distribution* of events in a large sequence of throws. Likewise, in the above case in which we don't have perfect knowledge of the initial conditions we will be faced with uncertainty of this epistemic type: there will be a statistical spread of possible outcomes. This links the discussion to probabilities and their interpretation.

There are three broad categories of interpretation regarding probabilities:

- **Objectivist.** Probabilities pick out 'real' features in the world, independently of the existence of humans. The *relative frequency* interpretation according to which probabilities are ratios of repeated events views probabilities objectively in terms of a correspondence between the probability and the number of times (or percentage) an outcome is found in the repeated run (strictly speaking an *infinite* run). It is possible to think of objective probabilities in terms of 'propensities,' which are dispositions[13] to produce the kinds of outcomes that would be able to *ground* the kinds of relative frequency just mentioned.
- **Subjectivist.** Probabilities refer to the degree of belief (a value between 0 and 1) that an agent has. It is, of course, dependent on the agent's beliefs and is sometimes called the 'personalist' approach since each *agent* determines their own interpretation of probabilities. This is tamed (made more 'objective') by adding various kinds of constraints so that agents are forced to at least be consistent in their assignments of probabilities. Note that probabilities for single events can be dealt with on this approach (e.g. where we do not have the luxury of an ensemble of copies of the event).

- **Evidentialist.** Probabilities are objective facts about the levels of support between empirical claims. This is a *inductive logic* approach analyzing relations between statements: it locates probability neither in the world nor in the head, but in a kind of abstract formal space.

These categories split apart into sub-categories; however, we need not concern ourselves with the details here. What matters is that some treat probability as a feature of the world ('ontic': the world itself is 'chancy') while others treat it as a mental contribution ('epistemic': the world may or may not be chancy, but our knowledge of it is uncertain). To return to the case of the toss of a die above, it might be that there is a fact of the matter, determined by physical law, about which side of the die will be face up, the subjectivist can still assign a probability based on incomplete knowledge of the situation.

In the case of quantum mechanics, the orthodox interpretation is that there is 'ontic uncertainty' (objective probability) since the laws are themselves only capable of generating probabilities for outcomes. As we see in Chapter 7, quantum mechanics caused many to give up on the notion of determinism, and the related notion of cause and effect. This reflects the 'standard viewpoint,' one replicated across countless popular TV programmes on physics (but also more serious academic literature), that classical physics is deterministic and quantum mechanics came along and destroyed this neat deterministic picture, underwritten by the uncertainty principle (sometimes called 'the principle of *indeterminacy*'). However, we need to be far more careful in our assessments of whether a theory is or is not deterministic – and after all, we are concerned with *physical theories* here (and their interpretation), and whether we think that our world is *actually* deterministic will depend on what our best theories say, and how faithfully we take them to map to our world.

Defining Determinism

We also need to be careful in how determinism is defined: what exactly are the necessary components of this thesis? We saw that it is usually bundled together with causality and prediction, but this has recently come under fire. When we pull apart these elements, we are left with a formulation that forces us to revise the standard view. Causality faces the troubles identified by David Hume long ago, and solidified by Bertrand Russell: causation does not appear in our theories; rather, all we have are functional relationships of various kinds. Causation faces too many philosophical problems of its own to make it reasonable to base a definition of determinism on it. Likewise prediction, which also fails to secure an ontological notion of determinism since to predict is to perform a mental act, and this is highly dependent on what skills we attribute to whatever is performing such acts. True, if we can make accurate predictions using some theory then it perhaps offers up some evidence toward that theory's status in terms of determinism, but strictly speaking it belongs in the realm of epistemology. Moreover, the existence of chaos, which involves a lack of predictability yet still, we want to say, is deterministic, should also lead us to wish to tease these two concepts apart. Again, since theories are our guideposts to reality, we ought to couch our definition of determinism is terms of theories and their interpretations.

The preferred formulation of modern philosophers of physics can be discerned from some of what we already said above. Let's assume that a theory is defined by the *states* and *laws* governing its systems of interest. The trick is then to consider *pairs* of systems that are 'prepared' in the same way (i.e. in the same state) at some instant of time. Given such preparation, determinism is the claim that the systems will share the same state at all future times, so long as they are subject to the same laws:

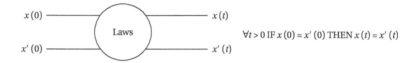

This makes no mention of predictability and causality, though they can easily be incorporated. What's more, this account can easily be extended to the consideration of entire histories (i.e. universes or worlds): determinism just means that worlds that agree up to some time agree at all times. The hole argument can be considered in just this way by thinking about the state of an entire (instantaneous) slice through the spacetime (the universe) and all of spacetime before that slice and asking whether the future behavior of the fields defined on the slice (from which the state is constructed) are uniquely determined. Couched in our preferred way of thinking about determinism, we then ask whether it is possible to have a second universe identical to the first up to this same slice but differing thereafter. Of course, we found that it was indeed possible to have a second universe provided we treated the spaces' points as real entities independently of the fields defined with respect to them: the symmetry of the theory means that we can deform the fields to the future of the slice while still producing a legal solution of the equations (i.e. while remaining consistent with the laws). But this same freedom means that the very slicing we used, to set up the test of determinism, is itself unphysical. (If we are thinking in terms of worlds – or rather *models* of worlds – instead, then the identities between initial and final states above would be instead isomorphisms between [portions of] the worlds.)

Denying Determinism

Several notable violations of determinism occur as a result of 'interference' or a breakdown of the theory (so that a solution cannot be extended to later times). For example, a major problem that threatens determinism in a Newtonian universe (the natural environment for Laplace's demon's party-trick) is the absence of any speed limit. Causal influences can propagate at whatever speed you like. Interactions can be infinitely rapid: indeed gravitation is a perfect example of such an influence, though not one mediated by a propagating particle of course. This implies that particles are able to shoot off to or in from 'spatial infinity' in a finite interval of time, so that they don't show up asymptotically in the space, or show up without being in the space at the start of the interval – the inward particles have been dubbed 'space invaders'! This allows for future states to be meddled with in a way not determined by an initial state. Perhaps surprisingly, special relativity makes the world safer for determinism, since the speed limit (involving a transformation of causal structure of spacetime) prohibits such space invaders and their reversals. Another option for outlawing space invaders and defectors is to enforce (global) conservation of energy: after all, particles coming in and out of the world will be bringing (creating) and taking (destroying) energy as they do so – this also raises the point that the number of particles in the world is not invariant, which might be a cause for concern.

Another failure of determinism in Newtonian mechanics concerns the simple breakdown of the applicability of the theory as a result of a (collision) singularity that occurs because of the form of the inverse-square law. This of course contains a $1/r^2$ term, where the r is the distance between a pair of particles. If we consider the mutual gravitational attraction of a pair of particles, a collision will obviously mean that the distance is zero. The laws of the theory simply cannot determine what will occur after such a singularity has occurred.

An example of Newtonian indeterminism, that puts us in mind of Zeno's paradoxes, was devised by Jon Pérez Laraudogoitia [35]. Known as a 'supertask' (performing infinitely many steps in a finite time), it goes as follows. Firstly, we need infinitely many point masses, arranged along a meter-long line spaced according to the infinite geometric series $1, \frac{1}{2}, \frac{1}{4}, \frac{1}{8}, \frac{1}{16}, \ldots$. The first particle, at the start, is taken to be moving toward the second particle at $\frac{1}{2}$, which is then pushed onto the remaining particles, one after another, all at a rate of one meter per second – obviously, since this is laid out in a 1-meter line, the whole thing will be over in a second. During each (elastic) collision a particle p_n will transfer its momentum to p_{n+1}, thereupon coming to a state of complete rest where p_{n+1} was previously at rest – one can envisage a version of the toy known as 'Newton's cradle' with infinitely many balls. After a second all the collisions will have completed, and the entire system will be at rest. But, and this is where the indeterminism springs from, if this is a possible Newtonian process (as it appears to be), then so is its time reverse (since Newtonian mechanics does not have a preferred direction of time). If we play the tape backwards in this case we have what appears to be a *spontaneous* self-excitation of the particles' motion at $t > 0$, which of course conflicts with determinism. What is curious about this example is that the momentum (in this case the energy $\frac{1}{2}mv^2$) that we imparted at the start has been gobbled up by an infinite sequence. This again points to non-conservation of momentum (at least at a global level).[14]

General relativity is far more complex to deal with, and we see clearly the context-dependence of determinism in the fact that the particular spacetime structure is based on the particular solution to the field equations of the theory. There is simply too much freedom in creating universes – though this provides a theme-park experience for philosophers of physics. Virtually all of the troubles stem from the fact that general relativity's equations are 'local' in that they link curvature and energy at a point. While locally things look quite simple (approximating Minkowski spacetime), globally things can become unhinged in a variety of ways: the global structure is something to be fixed by hand rather than by the theory. Some of these choices (i.e. for some choices of energy distribution) are better suited to determinism than others; some don't even allow for the setting up of the initial value problem in which determinism is couched (i.e. data on an instantaneous slice that is 'pushed along' by the laws, thus generating the spacetime: a solution to Einstein's equations). Spaces with 'closed timelike curves' (theoretically, those permitting time travel) are of this kind. In general relativity, given the dynamical nature of spacetime (coupled to mass and energy), an infinitely dense mass creates infinite curvature, which effectively creates an 'edge' to spacetime: a singularity.[15] This is a generic feature of the worlds of general relativity. The existence of singularities in general relativity leads to the problem that one will have situations in which the theory cannot predict what will occur at such singularities (as with the Newtonian singularities above). One way of viewing this breakdown of determinism is in terms of a *limit* of the theory's applicability, pointing to some successor theory able to deal with the singular behavior, or able to smear it out somehow.

There are more arbitrary ways in which determinism can be made to break down in general relativity, for example by simply 'deleting' that part to the future of some slice through spacetime (and that slice itself) in which case we have an abrupt *end* to the spacetime. This, strange though it may seem, is a physically possible world according to general relativity. While not very satisfying, we can clearly see that handling the issue of determinism in general relativity is fraught with difficulties and exotic potential counterexamples.[16]

As we will see in Chapter 7, quantum mechanics is not necessarily indeterministic: so long as a kind of nonlocality is preserved in the quantum theory it is perfectly possible to have a deterministic version (known as de Broglie–Bohm theory). The infamous many-worlds interpretation is also deterministic in that the total state (represented by a wavefunction for the entire universe – or, rather, 'multiverse') at any time suffices to determine it for all times – what is problematic, however, is making sense of *outcomes* and their probabilities for realization in a world in which 'everything happens.'

5.4 Further Readings

As with the previous chapter, many of the discussions of the topics in the present chapter lie more within metaphysics and other areas (such as the philosophy of time, chance, and probability) than philosophy of physics.

Fun

- Craig Callender and Ralph Edney (2001) *Introducing Time: A Graphic Guide.* Icon Books.

  - An excellent overview, in brief cartoons, of many of the major philosophical topics in philosophy of time (dealing mostly with physics-based issues).

Serious

- Barry Dainton (2010) *Time and Space* (2nd edn). Acumen Publishing.

  - A very clear and comprehensive treatment of issues in the philosophy of space and time, including both philosophy of physics and more metaphysical issues.

- Lawrence Sklar (1974) *Space, Time, and Spacetime.* University of California Press.

  - Slightly older, but still comprehensive introduction to issues in the philosophy of spacetime physics. It covers the epistemology of geometry in great depth, and also covers the relationship between causal ordering and time (not covered in this chapter).

Connoisseurs

- John Earman (1986) *A Primer on Determinism*. Dordrecht: Reidel.
  - The classic text that did much to modify the simplistic discussions of determinism in classical and quantum theories. Its 'connoisseur' level placement is not an indicator of its reading difficulty: it is a sparkling read, as with his book *World Enough and Space-Time*.

6 Linking Micro to Macro

Thermal physics has had a tendency to leave philosophers of physics rather . . . cold. Not quite as mind bending as quantum mechanics and relativity, it has been somewhat neglected. Perhaps this is because thermodynamics began life as part of an engineering problem concerned with extracting the maximum amount of work from fuel, and thus with the efficiency of engines? Statistical mechanics 'grounded' these engineering results in a theoretical framework describing the constituents of complex systems exhibiting thermodynamic behavior. However, at the same time, statistical mechanics sits at a 'higher level' than, e.g. classical or quantum mechanics, since, to a large extent, it transcends the constitution of the matter involved: any system with macro-properties generated by micro-properties is potentially an object of study for the statistical mechanic. In this sense it resembles a *principle* (like evolution by natural selection) rather more than your typical theory of physics.

However, as a challenge for philosophers of physics, and a rich source of unresolved problems, it can hold its head up high – and fortunately, more recent years have seen a dramatic rise of 'philosophy of statistical physics.' The kinds of issues one finds are arguably the most interesting from a philosophy of science point of view: questions of emergence and reduction, and causality and explanation are a central part of the interpretive enterprise. However, though born in the study of heat, thermodynamics and statistical physics has also provided one of the most perplexing puzzles about time that we know of: a time-asymmetric law at the heart of physics!

This chapter begins with a brief account of thermodynamics, based around the study of heat engines, to express the basic laws of thermodynamics, and introduces some basic concepts (efficiency, entropy, randomness, reversibility, etc.). Standard topics involving reduction (to statistical mechanics), interpretations of probability, Maxwell's Demon, and time asymmetry also are then introduced. Discussed too are topics that have only recently made it into the mainstream philosophical literature, such as the links to cosmological issues. The puzzle of organization in our universe (given the second law) is discussed in considerable detail, including the puzzle of the low-entropy past, and Boltzmann brains (and the notions of selection bias and 'Anthropics' in physics).

6.1 Thermodynamics, Statistical Mechanics, and Reduction

Thermodynamics, as the name suggests, is the study of the thermal properties of matter ('the scientific study of heat'). Statistical mechanics is a study of complex systems – i.e. reasonably sized objects with many parts. But rather than studying such objects by decomposing them into their simpler parts, and studying those in order to learn about the complex system, the subject deals with more coarse grained ('equilibrium' or simply 'thermodynamic') variables. Common examples are temperature, mass, pressure, volume, and entropy. In a sense, these variables bundle together the properties of the constituent parts (whatever they may be) into a single number characterizing the whole system. But we don't worry about what the simpler parts are doing, or what they are made of. Thermodynamic processes link pairs of states (initial and final), characterized by distinct thermodynamic variables (e.g. hot to cold), and satisfying boundary conditions (describing constraints on how the process may unfold so that not all formal possibilities represent physical possibilities).

In the earliest days of thermodynamics, the properties and laws governing systems were viewed from a fairly high 'phenomenological' level, without much concern for the underlying constitution and mechanics. As mentioned in the introduction, the aim was to study the relationships between various thermodynamic quantities and processes that utilize them with a view to getting the *most* work from the *least energy* (i.e. maximum efficiency so that as little useable energy as possible is wasted). Yet most of the central concepts and laws of thermodynamics were discovered without the benefit of a microscopic explanation of *why* they hold. The key idea of statistical mechanics is that the coarse thermodynamic properties and laws can be reduced to laws concerning the behavior of micro-particles. As we will see, however, this reduction is bristling with conceptual curiosities.

Statistical mechanics is a set of tools that enables us to discuss the 'macroscopic' thermal features of the world by looking at the mechanics of the 'microscopic' constituent parts – generally, "micro-" and "macro-" do not pick out definite scales, but *levels of analysis*, so that the microscopic parts might well be visible to the naked eye (in which case, the macro-system would be significantly larger). Think of a large musical ensemble playing in concert. There is a 'global' sound, which is built from the behavior of the 'local' degrees of freedom (the various musicians). In the case of statistical mechanics, the musical output is like the temperature or pressure of some gas and the musicians are like the molecules comprising the gas – this is of course just a rough illustration: I'm not suggesting that statistical mechanics is applicable to systems with so few constituents. One can analyze the music from this global level, looking at the broad contours of the music: its melodic structure, musical form, and so on. Or, alternatively, one can focus in on what the individual musicians are doing to generate this well-coordinated global output. It is this 'zooming in and out,' from the whole system to its parts, that characterizes the relationship between statistical mechanics and thermodynamics (and, more generally, the relationship between statistical physics and the more coarse-grained, observable features of systems): you won't see harmonies in a single bassoon; likewise, you won't see pressure or temperature in an individual molecule. (Though we won't say so much about it, there is often something *novel* that comes with 'collectives,' new kinds of order and laws that are not shared by the parts: "more is different" as the physicist Philip Anderson expressed it.)

The central task facing the statistical mechanic is, then, to try and recover known thermodynamic (and other) macro-properties from the behavior of individual parts. Of course, there are very many parts to analyze in the case of the molecular structure of gases and other macroscopic objects, so a direct enumerative approach, in which one simply figures out what each individual is doing (e.g. by integrating the equations of motion one by one), will not work. One needs to employ a statistical approach (and therefore probabilities) because of this enormous complexity in organization. The macroscopic thermal properties emerge as statistical phenomena: averages over properties of the individual micro-constituents – these particles are understood to be perfectly well-behaved, but there might be 10^{23} or 10^{24} of them, each with their own trajectory to be solved for. In this case, the postulates of the theory of gases and heat become theorems of the statistical approach. However, as the label 'statistical' suggests, the *theorems* have the status of truths only 'on the average': for all practical purposes, they can be taken to be true since exceptions are incredibly improbable.

Already in the very act of defining statistical mechanics and thermodynamics we have introduced two deep issues:

1   How are the probabilities in this approach to be understood?
2   What is the precise nature of the relationship between the micro- and macro-levels?

With regard to the first matter, there is an oddity in the fact that probabilities are appearing at all: after all, classical physics does not include any in-built indeterminacy (of the kind found in quantum mechanics). As the example of Laplace's demon makes perfectly clear, given a specification of the positions and momenta of a system of particles, the laws of classical mechanics enable one to generate, by integration, the *unique* future evolution of the system. Yet statistical mechanics renders the statements of classical thermodynamics into probabilistic ones. At the most basic level, as with most situations where statistics must be resorted to, it is the sheer computational complexity (and therefore impracticality) that forces the introduction of probabilities.

In response to this puzzle of how such probabilities ought to be interpreted, one usually finds adopted a *frequentist* stance based on the mathematical result known as 'the law of large numbers' (or, more commonly, the law of averages). The idea involves first defining a probability as the value of a distribution function for some random variable that we view as representing the experiment we're interested in (e.g. a simple toss of an unbiased coin, with outcomes 'H' and 'T'). Here one is interested in the random variable H_n/N (the average value telling us the fraction of outcomes that are heads), where N is the number of coin tosses in the trial, and H_n is the number of times the outcome lands heads. This will clearly take values between 0 and 1, and as N gets larger, H_n/N gets closer to 0.5, such that $\lim_{N \to \infty} \frac{H_n}{N} = 0.5$.

The reasoning is quite intuitive: the frequencies here involve various possible arrangements of things (microstates) such that the number of these various arrangements (some of which will be effectively identical) become probabilities. Moreover, if we don't have some 'inside knowledge' about these microstates, then it makes sense to assign equal probabilities to each distinct possibility. Hence, we give the probabilities an *ignorance* interpretation.

We will see that this is quite unlike the probabilities we find in quantum mechanics. There is nothing inherently probabilistic about the motions of any of the particles: given a specification of the initial values of the positions and momenta of the particles, their future trajectories are uniquely determined: Laplace's demon would have no need of statistical mechanics. We do need it because of our *ignorance* of these instantaneous values, so we average over their values – this notion of an average value was the original meaning of a 'statistic,' of course. We face a problem here in that there are no probabilities as far as the parts are concerned, but probabilistic notions characterize the complex systems they form when bundled together.

The second issue also has a similar oddity in that along with the probabilistic features that occur, despite the deterministic components, there are also other features that point to a mismatch between the levels: the macro-level, for example, involves time-asymmetric phenomena while the microlevel does not. Hence, the reversibility of the particles appears to be washed away by the statistical description. This is one of the key puzzles we discuss in what follows.

6.2 Approaching Equilibrium

One of the worst things to face in one's life is when someone (or yourself) breaks wind in a confined space, such as an elevator. We know what the dreaded outcome will be. Though the nasty molecules will commence their life in a well localized area (you know where. . .), they quickly spread to wreak havoc elsewhere, until eventually they occupy the entire elevator and, thank heavens, are diluted enough not too cause anymore bother. This is wholly unsurprising behavior for a gas, and is well described by thermodynamics – though I expect that this particular example might not come up in many thermodynamics exams. We will meet this idea again, in §6.3, when discussing Clausius' treatment of entropy and the second law.

We can represent the general process of the transition from an initial equilibrium (where there are constant values for various properties like heat, pressure, and so on) to a breaking of equilibrium (an introduction of heat, pressure, or some other disturbance to the system) to a restoration of equilibrium once again as follows using the standard example of a box with a partition wall through the center, with the gas initially confined to one side. The gas is initially in equilibrium, but then we remove the partition wall (an intervention), which, of course, results in dissipation. Eventually the spreading and mixing with other molecules is maximal (i.e. random) and the gas settles into its new equilibrium state (see fig. 6.1).

The example of breaking wind in an elevator has some surprising philosophical implications. Crucially, it constitutes an apparently *irreversible* process: a process with an arrow of time. Alas, one will never witness a time-reversed situation in which the odour clusters together once again near the culprit (as in the the process represented in fig. 6.2). In the case of our more orthodox, generic box of gas, once we have reached the new equilibrium, then, without some disturbance (such as some restoring force

Fig. 6.1   The approach to equilibrium: a gas started in an asymmetrical (low-entropy) state will tend toward a state of maximal equilibrium.

Fig. 6.2   The reverse of the process in fig. 6.1 – from equilibrium to low-entropy – does not seem to occur in nature.

pushing the molecules back over to their starting point on the left side), the reverse process restoring the initial state won't *spontaneously* occur without intervention – thus 'opening' a 'closed system.'

Perhaps looking at boxes of various shades of grey doesn't look so amazing. But this same irreversibility is everywhere. Many's the time I've left a cup of tea to brew too long and go cold. How nice it would be to be able to reverse the entire process and start again. But think what it would involve. The tea, water, and milk must be unmixed. The milk and tea must be cooled down, and the water heated up. Any evaporated water and milk must be recaptured. It is perfectly conceivable that one could do some clever manipulation (say with centrifuges, heating coils, refrigerators, and the like) to perform such a feat. But one cannot readily conceive of the time-reverse trajectory of the actual process occurring spontaneously. It's a nice 'exercise for the reader' to look around and see how many other such irreversible processes you can find (that is, processes that we never seem to experience happening in reverse temporal order): melting ice; steam from a kettle; cracking eggs into a pan; wood burning away in the fireplace; the sound coming out of your speakers or headphones; light from a lamp; jumping from a diving board into a swimming pool ... (Remember that it is crucial that the reversed trajectory must be driven by the natural dynamics of the system as it is in the usual 'forward direction': it must 'just happen.'

This irreversibility is part and parcel of the *second law of thermodynamics*: the natural order of things (i.e. without intervening) sees order (low entropy) go to disorder (high entropy), rather than the other way around. In order to find an explanation for a law or property, one usually looks to lower levels: here the level of microscopic constituents. In the case of the gas in the box, these are molecules. This is the realm of statistical mechanics of course. But, and here's the rub: the laws (e.g. Newton's laws, or possibly the laws of quantum mechanics) governing the microscopic constituents *do not experience any arrow of time*. For any microscopic process going in one direction of time, there is a possible process (not in violation of the laws of physics) going in the other, reverse direction of time – this simply corresponds to the mapping, $t \mapsto -t$. In other words, the laws of the particles are time-symmetric (something that applies to both classical and quantum mechanics) and therefore insensitive to the observable arrow of time at the macroscopic level. (I should perhaps mention that weakly interacting particles, e.g. B mesons, *do* appear to violate time symmetry, though the combined *TCP* operation of time reversal T, along with parity inversion P [mapping a particle to its mirror image] and charge conjugation C [mapping a particle to its anti-particle] see symmetry restored. However, this violation of time symmetry does not point to any kind of arrow of time, in the sense of *irreversibility*, since the reverse process is a possible process:

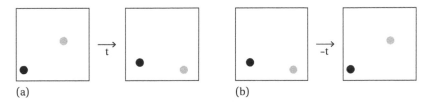

(a)   (b)

Fig. 6.3   Worldlines of two particles in relative motion where (b) is the time-reverse of (a).

it simply exhibits slightly different properties from the original temporal orientation, such as distinct decay rates.)

The simplest way to explain this time symmetry is to invoke a movie of a bunch of atoms colliding like billiard balls. Watch such a movie for ten seconds, with the atoms going in their various directions. Then play the movie backwards. Both would be perfectly acceptable time evolutions for such a system (i.e. they are possible solutions of the equations of motion). For example, in figure 6.3 we see two movies with the same two particles. In the first movie (*a*), the dark particle moves up and to the right, just a small distance, while the light particle moves down and to the right, more quickly (covering more distance). In the second movie (*b*), the particles begin in the final position from movie (*a*) and end in the initial position of movie (*a*). In terms of laws, this amounts to inputting (*a*)'s final state as an 'initial condition' into the mechanical laws and generating (*u*)'s initial state as an output (this can be achieved by changing the signs of the particles' velocities).

If we were to focus entirely on the particle worldlines for this ten second period, we would see that they were identical – see fig. 6.4. The form of Newton's third law of motion, $\mathbf{F}(\mathbf{x}) = m\frac{d^2\mathbf{x}}{dt^2}$, makes this self-evident since the second time derivative doesn't care whether we use t or $-t$ – note that dependence on the second time derivative is not a necessary condition for time-reversal invariance; it just happens to be responsible in this Newtonian example. This means that if $\mathbf{x}(t)$ is a solution (corresponding to the trajectory shown in movie (*a*), then so is $\mathbf{x}(-t)$ (corresponding to

Fig. 6.4   Worldlines of the two particles from the previous movies, which you can think of as a photographs with a ten-second exposure time. Naturally, this picture would be the same for t and $-t$ cases.

the trajectory shown in movie (*b*), with the physical solution given by the equivalence class of time-symmetric solutions, as sketched in fig. 6.4.

So the challenge for the theory is to find a way of getting time-asymmetric processes out of the time-symmetric laws, the former somehow emerging from the latter. Putting it in terms from above: how do we get time-asymmetric macromovies from time-symmetric micromovies? If it is possible (in terms of the microlaws) for all of the particles in a gas to undergo inverse motions (e.g. by flipping the signs of particle velocities), of a kind corresponding to the time-reverse of the gas in the box (or elevator!), then why do we not observe such things happening? From whence the arrow of time?

Ludwig Boltzmann attempted to resolve these problems after they were raised by Josef Loschmidt – the point is known as "Loschmidt's Objection." The crux of it is that given the time-symmetric nature of the laws governing the particles making up macroscopic systems, we ought not to see time asymmetric behavior. Entropy decreasing jumps, such as spontaneous coffee/milk unstirring should be observable. As philosopher Huw Price puts it: symmetry in, symmetry out. Conversely, to get an asymmetry out demands an asymmetry in the input. This input asymmetry, as discovered by barrister-turned-physicist Samuel Burbury, is the assumption of *molecular chaos*: pre-collision particles (i.e. their velocities and positions) are uncorrelated (probabilistically independent: random), but not so post-collision. In other words, the collisions of the particles are responsible for the observed entropy increase such that for any given distribution of states of the particles, it will inevitably evolve into an equilibrium state (the maximally random Maxwell state). Of course this leaves a problem that we will turn to in §6.7: if the system is already in a high entropy state, then it has nowhere to go. One needs some asymmetry in the dynamical evolution, but also an asymmetry in the boundary conditions: one needs lower entropy in the past so that the dynamical asymmetry can generate its effect.

6.3 The Laws of Thermodynamics

I hope you're drinking a nice cup of tea while reading this book. When you pick up that cup (if you don't have one, perhaps make yourself one), you are doing *work* of course – battling against the forces of gravity. The work converts kinetic to potential energy as you hold it up, which if (don't try this bit) you should let go of the cup, will be converted into kinetic energy as the cup falls to the floor, shattering and spilling the lovely tea everywhere. Where does the energy go once it is lying in a mess on the floor? It went into thermal energy of the pieces and tea, and the floor and the air – of course, some of the energy already went toward heating up the air molecules in the cup's path on the way down, subjecting it to air resistance. This is an example of the so-called *first law of thermodynamics*: total energy is conserved in a closed system (roughly, the room in which the cup was dropped immediately after it was dropped). The energy in this case was transformed between various forms: potential → kinetic → thermal – where the latter is, once analyzed, simply another form of kinetic energy (namely, that of the particles). Pre- and post-process (cup dropping) energy is a conserved (invariant) quantity.

There are certainly some interesting issues surrounding the first law, not least connected to the issue of 'open' versus 'closed' systems, which it involves. However, the real philosophical meat is to be found in the second law.

The *second law* can be seen in the same example: whereas initially the energy could do work – e.g. imagine a pulley system in which one end of the rope is attached to the cup and the other to a small mass (sitting on the floor perhaps) weighing less than the cup, so that dropping the cup from the table would see the mass rise – afterwards, when there is only thermal (random) energy, one cannot do useful work anymore, despite no loss of total energy in the whole system. Initially, there was a cooperation on the part of the molecules making up the cup and tea within it; a team effort of motion toward the floor. In other words: order. Afterwards, once it has come to rest, randomness prevails, with the molecules whizzing in all directions. This would be an equilibrium state – of course, equilibrium here does not mean some kind of zen-like 'still point'

The concept of entropy is at the root of the second law. Rudolf Clausius, building on Sadi Carnot's work on the efficiency of steam engines, introduced this concept and with it the law to which it is associated. Clausius was interested in the movement of heat (thermodynamics), and noticed that it has a curious uni-directionality: objects initially at *different* temperatures when placed together so they are touching, will approach the *same* temperature. But this process of balancing has an endpoint at which there is no more flowing of heat across from the hot body (or gas) to the cold one: the heat gradient that existed earlier is no longer there. This never happens in reverse: hence, the second law – in this case, a matter of finding a balance so that the temperature is uniform (that is, at equilibrium). 'Entropy' was Clausius' term for this tendency.

As with the first law, we need to include a statement about 'closed' versus 'open' systems, since clearly we can sometimes heat things up (the tea in your cup, for example) and cool things down (the milk for your tea from the refrigerator). Both require the input of energy, and the energy must exist in a low-entropy form (or lower than the system to which it is introduced) in order to perform work.

This takes us back to the first law, which, though it demands conservation of energy, does not forbid it from taking different forms. Some of these forms are more capable of doing work than others, and these are precisely those with the lowest entropy. Going back to the pre- and post-broken teacup, we can see that the pre-broken cup was in a lower entropy state than the post-broken cup since we could do some work with it (e.g. lift some object, as mentioned earlier, or use its more concentrated warmth to heat something).

These laws are of a similar kind in that they involve comparing quantities before and after some process has occurred. However, in the case of the first law, there is an invariance of a quantity before and after, in the case of the second law there is an inequality (always involving a quantity being larger than before). The quantity in the latter case is entropy S, to which we turn next. We also find that the second law, though it appears *exact*, was later understood to be probabilistic (i.e. true in the vast majority of situations). The concept of entropy is modified accordingly, and becomes a statistical concept (roughly having to do with the number of ways a state can be realized by its parts).

6.4 Coarse Graining and Configuration Counting

Our modern notion of entropy comes from Ludwig Boltzmann (superseding the more phenomenological version due to Clausius, relating entropy to energy and work). In a nutshell, Boltzmann's idea is that entropy has to do with counting possible states for a system. However, we must split the types of possible state into two families, depending on whether they refer to the level of particles or the level of the systems made up of lots of particles. High entropy then simply means there are lots of ways for some state to occur (at the level of individual particles), and low entropy means that there are very few ways for some state to occur (at the level of individual particles).

The definition involves a mathematical procedure known as 'coarse graining,' which, as the name suggests, neglects certain fine but observationally irrelevant details of macroscopic properties. Starting with a system's phase space, one divides it into cells, such that all of the phase points within a cell (henceforth: 'microstates') correspond to one and the same macroscopic state (henceforth: 'macrostate'). In other words, the cells are equivalence classes where the relation of equivalence is 'has the same macrostate as,' which just means that one would not be able to distinguish between those microstates by observing the macrostate. An assumption of equiprobability is then made, such that each microstate is just as likely as any other. Crucially, this does not mean that each macrostate is equally as likely to be realized as any other: those macrostates that are associated with more microstates clearly have an advantage.

Let's bring in a few technical details here, which aren't so difficult – some of which we encountered in earlier chapters. Firstly, recall that the phase space (i.e. the state space) is understood to represent the space of possibilities for a system, here of *n* (indistinguishable) particles – indistinguishable, but still distinct and countable (giving the so-called Maxwell–Boltzmann distribution). Since it includes information on both the positions *q* and momenta *p* (for each particle) it will be a 6*n* dimensional space, which we call Γ. The idea of partitioning the space into cells is that while all points *x* = (*q*, *p*) represent physical possibilities, some points are to be seen as representing, many-to-one, *the same* possibility (in terms of macroscopically observable quantities): the physically measurable quantities one is interested in will be insensitive to differences amounting to swapping any such points. One can then think of the cells as themselves being the mathematical entities that correspond one-to-one with macrostates, so that each cell corresponds to a macroscopically distinguishable feature of the world. We can label each cell by coordinates w_p and let us also represent the volume of each cell by W_i (see fig. 6.5).

This small amount of apparatus allows us to now introduce Boltzmann's famous equation for the entropy *S* of system in state *x*, where $x \in w$ (that is, the state *x* of the system lies within the cell *w*). It simply tells us that the entropy is the natural logarithm of the volume of the phase cell in which it sits.

$$S = k \log W \tag{6.1}$$

It is then easy to see that the larger the volume *W* of the cell *w* that the system's state *x* lies within, the more ways there are to 'get' that state; the higher the entropy will be; and (Boltzmann argued), the more likely it will be to occur.

Putting it back in terms of micro- and macrostates: macrostates realizable in more ways have a higher entropy and are more likely. For example, *W* tells us how often we can expect some temperature (of say 30° Celsius) to be realized. (Note that the number *k* is Boltzmann's (conversion) constant used simply to get the two sides to agree in terms of units: $k = 1.38 \times 10^{-23} J(\text{oules})K(\text{elvin})^{-1}$. It can be viewed as a bridge, linking microscopic properties of individual particles (energies) with macroscopic parameters, such as temperature. The logarithm is used to get a grip on the huge numbers of microstates that some macrostates will be realizable by.) The largest cell will be that corresponding to the (thermal) equilibrium state, with the most microstates. This has the largest entropy. Hence, in much the same way as it is overwhelmingly likely to pick a red ball from an urn with 99 red balls and one black ball, so it is overwhelmingly likely to find the above system in a larger-volumed cell (assuming, as indicated above, that all of the microstates have an equal chance of being realized). The probability of some state is simply given by its volume (viz. the extension in phase space). The most probable state is then that realizable in

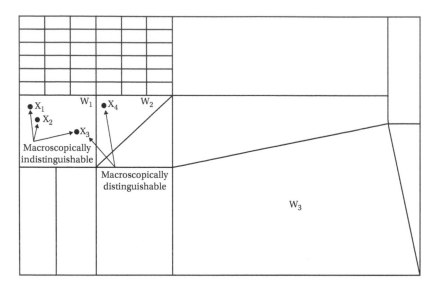

Fig. 6.5   An exaggerated example of a coarse graining of phase space Γ into cells w_i of volume W_i. The phase points (microstates) x_1, x_2, and x_3 in cell w_1 are indistinguishable (correspond to the same macrostate), while these points and x_4 in cell w_2 are distinguishable. Cell w_3 has the largest volume (i.e. the greatest number of microstates capable of realizing it). It can be seen that $W_3 > W_1 > W_2$. The geometry of phase space tends to make it overwhelmingly likely that the system will be in the cell w_3 where it will spend most of its time.

the most ways – this was called the *Maxwell state* by Boltzmann, since it conforms to a distribution discovered by James Clark Maxwell. This is the state that will found to occur most frequently and allows one to recover the (previously mechanically inexplicable) second law:

$$\frac{dS}{dt} \geq 0 \tag{6.2}$$

But it is, of course, a *statistical* explanation: it is probabilistically extremely likely to see the second law satisfied by the kinds of processes we are capable of observing because we are dealing with such large numbers of particles, which makes it vastly more likely that they don't spend time in macrostates with less microstates.

Now, though this is rather elegant and easy to understand, there are all sorts of assumptions made, and there is an inescapable vagueness in the notion of "macroscopic indistinguishability." One might easily be disturbed by the apparent looseness in doing the coarse graining: how is the carving up of the space carried out? How do we know that my carving will be the same as yours? Much depends on pragmatics: what level of *precision* is required? If you need to distinguish states more precisely than me, then you will use a finer 'mesh' size to cast into phase space, trawling for classes of indistinguishable states at that grain. Moreover, given that the entropy involves the logarithm of the cell volume, our distinct carvings would have to differ radically to get appreciable differences for *S*.

However, this understanding of entropy also has important consequences for some of the puzzles about asymmetry in time. The basic idea is that the number of ways to be in a 'disordered' (high entropy) state is greater than the number of ways to be in an 'ordered' (low-entropy) state (the volumes of the former will be greater than the latter), so we should expect an increase in randomness. Indeed, Boltzmann argues that a closed system driven by random collisions of particles should tend to *maximum entropy* (such as occurs when our initially localized fart from above spreads maximally throughout the elevator). This result, known as 'the *H*-theorem,' accounts for the derivation of temporally asymmetric behavior from temporally symmetric microscopic behavior – '*H*' itself is a measure of deviation from maximal entropy, or the 'Maxwell state' (so that *H* is only found to get smaller over time). The probability for deviations away from the random Maxwell state will be smaller as the deviations become larger (by ever greater amounts). And, indeed, if one begins in a state that deviates significantly from maximal entropy, then it will have a large likelihood for approaching the higher entropy state. According to a common view, this is significant from the point of view of explaining the arrow of time: if the universe *started off* in a very improbable state (exceedingly far from maximum entropy), then it will have an exceedingly large likelihood of 'making its way back' to maximum entropy (via a sequence of higher-entropy states).

Let us now turn to the connection to such temporal matters.

6.5 Entropy, Time, and Statistics

The second law is probably the most well-known (and seemingly solid) law of nature. It says, quite simply, that entropy (for closed systems) never decreases – of course, one could 'open up' such a system and pump energy in, thus lowering the entropy (as happens when we clear the messy dishes and place them in the dish rack, for example). Much of the fun stuff in the philosophy of thermodynamics comes out of Ludwig Boltzmann's statistical (microscopic) explanations of thermodynamical behavior. Prior to the microscopic understanding he provided, the laws were understood in terms of the ability of systems to do *work* (such as powering a piston) or not. Entropy was then understood in terms of degrees of *irreversibility*.

Take my daughter Gaia's room, for example (see fig. 6.6). It is seemingly always in a state of near-maximal disorder. To impose order requires me to

Fig. 6.6   A messy bedroom: by far the likelier configuration. Relating this back to the coarse graining diagram, we could expect this to be represented by w_3, or one of the larger cells. Tidying the room 'pushes it' into one of the smaller cells. The second law then results in a motion toward the largest cell once again, so long as external energies (e.g. cleaning) are not put in.

input energy (cleaning it: in which case it isn't a closed system): hence, it becomes an 'open system' and does not 'naturally' tend to an orderly state. There are very few ways for it to be clean and orderly (with books organized alphabetically, and so on), and very (**very**) many ways for it to be disorganized. In terms of 'energy cost,' it is far cheaper to bundle books and clothes haphazardly than to have them arranged neatly – something Gaia knows all too well. . . Hence, there is a nice intuitive way of understanding entropy in terms of possibilities: it provides a measure of the number of possibilities for realizing some system state or configuration. Messy rooms can be realized in more ways than tidy rooms. It would not be a good strategy to randomly throw objects into positions: the number of tries it would take to 'fluke' the tidy configuration would be of the order of very many googol-plexes (which, in temporal terms would be very many times the age of the universe)!

   Aside: The 'intelligent design' brigade like to abuse this kind of reasoning to argue that the magnitude of order in the universe (particularly in the biosphere) can't be accidental since it's too improbable – like fluking the tidy bedroom configuration by randomly throwing objects about. So, they say, it must have been designed this way by a superior intelligence.

The problem (one of many. . .) with this argument is that the steps leading to 'the accident' would have to be uncorrelated (random) for this to work. But they are not random: evolution by natural selection is a law-based cumulative process involving vast timescales.

To repeat: Boltzmann's neat formula expresses this relationship between *the degree of disorder* (entropy, *S*) and *the number of ways that the system can be organized (or disorganized)*.[1]

This brings us back neatly to what is probably the most studied (by philosophers) aspect of thermodynamics: the appearance of an asymmetry of time, often called time's arrow. Our memories are of the past: the future is out of bounds. Yet, it appears that we have some degree of influence over the future, but not the past.

A common explanation for this asymmetry ties it to another asymmetry: that embodied in the second law. Since there seems to be an in-built irreversibility to the universe (entropy only increases and is irreversible for all practical purposes, with an unflinching march of order into disorder), the explanation for time asymmetry can 'piggy back' on this 'master arrow.'

Firstly, let's recall what is meant by "temporal asymmetry." What it indicates is that the laws (or phenomena described by some laws) are *not* time-reversal invariant. That is, the laws distinguish between a video of some process played forwards and backwards.

Think of the universe as rather like a grand version of my daughter's bedroom. We ought to expect, given Boltzmann's reasoning, that we should find it in a state of disorder (one of the larger cells). But, puzzlingly, we don't. We see an awful lot of order. Hence, from this perspective, our universe is very unlikely. Of course, we always knew there was something strange about it, but we can now quantify how odd it is.

Indeed, given our ordered state, we ought certainly to expect that the entropy goes up in the past, since there are more ways to be in high entropy states. This was the case according to Boltzmann's theory (which, according to him, 'boldly transcended experience'), who believed that the order we find is indeed highly improbable, but is bound to happen given unlimited time.

This explanation of time's arrow (by attaching it to the direction of increasing entropy) suggests that entropy-increase should be more probable both to the future *and* the past: there should still be a time-symmetry. The problem is, for some ordered state, it is more likely to have come from a less well-ordered state simply because those less ordered states far outweigh the ordered ones. Hence, Loschmidt's reversibility objection strikes again. We turn to Boltzmann's ingenious solution (invoking cosmology and the Anthropic principle) in a moment, but first let's consider another problem that faces the statistical account.

6.6 Maxwell's Demon

Recall that the second law states that entropy (in closed systems) is (for all practical purposes) unidirectional. In the second half of the nineteenth century the law was well known, and had already been tied to dire predictions about the fate of humanity: if useful mechanical energy steadily dissipates, then there'll be no power to run the world (and nature), with all energy eventually becoming thermal, random energy – you can't power anything with the ash from a fire. But recall also that the laws governing the particles' behavior are not necessarily unidirectional. Since the entropy changes are a result of the motion of the particles (i.e. it *supervenes* on the particles' properties), there seems to be no reason why the second law should hold (this is Loschmidt's objection again). By focusing in on the individual particle level, James Clerk Maxwell proposed a way of *beating* the second law (and the unhappy fate for humanity). It's a very simple setup, and really it doesn't demonstrate the falsity of the second law as such, but rather reveals its statistical nature – mentioned already, of course, but which found its earliest well-posed formulation in Maxwell's argument.

One imagines a box of gas, at some given uniform temperature, with the individual gas molecules buzzing around at different speeds (the average giving the temperature of the whole). Then one imagines placing a wall down the middle, dividing the box into two smaller boxes, and with a smaller hole in the partition that can be opened and closed at high speed (see fig. 6.7). We suppose that the two sides have the same number of molecules, and that they are at the same constant temperature. The two sides are at equilibrium relative to one another and in themselves. Though they are at equilibrium (so their mean temperatures are identical), the particles within will differ in speed at an individual level: some will be faster than others.

Next comes the demon, clearly based around Laplace's (in that it has access to the positions and momenta of all of the individual molecules in a system). It is really just a very high-speed sorting device that can keep track of the various particles whizzing around in a gas. The demon is able to allow fast moving particles through the hole into the left-hand side, say (quickly opening it as they approach), and keep the slow moving particles in the right (by quickly closing the hole). This sorts the two sides of the box into hotter and cooler gases, thus generating a temperature difference capable of doing work (such as driving a piston), yet apparently without doing much work to create it. The idea is that the demon apparently violates no laws of molecular physics with its actions, and yet has the effect of decreasing the entropy in the box. If this is a possible scenario, then it amounts to a perpetual motion device, since one is generating the energy version of a free lunch!

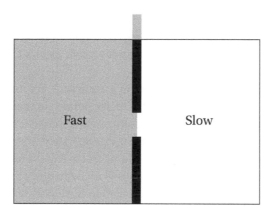

Fig. 6.7 Simple representation of Maxwell's Demon: by judiciously opening and shutting a door, a demon would be able to sort the particles so that hotter and colder regions are formed from a system initially in or close to equilibrium, in violation of the second law.

We all know there is no such thing as a free lunch, so to preserve the second law (statistical or not), we need to explain the apparent violation away. One way in which it could be preserved is if the missing entropy showed up somewhere else, such as the opening and closing of the shutter or the demon's own actions. Rolf Landauer famously argued that provided the demon recorded the measurements (with perfect precision) then the process he carried out would be perfectly reversible and therefore not entropy-increasing: if a record of some process exists, then the process is reversible; if not, then it is irreversible, and such irreversibility is the hallmark of entropy. The problem is, it would require a vast amount of memory resources to keep track of all those processes. In order to keep track, it would have to wipe its memory now and again. But this would make the process irreversible (leaving no record with which it could restore the earlier state), thus safeguarding the second law. Hence, Maxwell's demon needs to be absent-minded for the example to perform its function. However, wiping a memory creates entropy since the states held in the 'memory register' are shrunk down to one (0, say).

Other versions of Maxwell's demon come in the form of gravity, and more specifically via aspects of black holes. Black holes are defined by just three numbers: their mass, charge, and angular momentum. This enables a certain amount of 'forgetful' behavior that appears to offer a way to violate the second law. Simply throw a highly entropic object (such as the broken teacup) into the black hole and *voila*: the breakage never happened and the entropy is erased from our universe! However, this situation is avoided (and the second law saved) in virtue of a correspondence between entropy as usually conceived and the area of a black hole's event horizon:

the latter only ever *grows*, by analogy with the former. So when we toss in our broken teacup the horizon increases by a tiny amount, increasing the overall entropy in the universe – analogous versions of all of the laws of thermodynamics can be found.

Zermelo had objected to Boltzmann's statistical account of heat by invoking Poincaré's 'recurrence theorem.' He points out that given infinite time, there will be a perpetual cycling through all possible states, so that motion is periodic rather than settling into some fixed Maxwellian distribution. In other words, the distribution of states should be flat (with no preferred equilibrium value) since they will *all* be realized infinitely often. Again, at the root of the confusion appears to have been a breakdown of intuitions where very large numbers are concerned. When one compares the states that deviate from maximum entropy states with the maximum entropy state, one finds that there are vastly more ways of realizing the latter than the former. The problem Zermelo had is in understanding the fact that though the Maxwellian macrostates appear to be few, they preside over the greatest volume of microstates. By pointing to the statistical nature of the laws, Boltzmann brings statistical mechanics within the recurrence theorems. It is quite simply that any such recurrences are so improbable as to be rendered unobservable within any reasonable amount of time (what Boltzmann himself called a "comfortingly large" time). That is, the magnitude of the improbabilities means that they are ignorable for all practical purposes.

Of course, philosophers are not known for their adherence to practicalities, so we might still be rightly concerned by the conceptual implications of Boltzmann's view, for it indicates that we are simply around because of an improbable fluctuation in a vast bath of thermal energy that has existed for a very long time. Besides, Zermelo responded by pointing to a glaring hole (noted by Boltzmann himself) that we have no *physical explanation* for why the initial state (our ordered universe for example) has such a low entropy. Boltzmann thought this was akin to explaining why the laws of nature are as they are, and as such transcended physical explanation: one would be explaining the *grounds* of explanation.

6.7 The Past Hypothesis

Entropy is, to a rough approximation, disorder. It increases with time. That is a puzzle: why should it do that? But entropy was lower in the past than it is now. That is also a puzzle: why exactly is that? This is a quite different problem then. We have discussed the issue of why entropy *increases to the future*. But now: why does it *decrease* into the past? If there are more ways for our universe to be disordered rather than highly ordered, then why is it more highly ordered now? Why isn't the entropy much higher in the past as we move away from our present low-entropy condition (a question Boltzmann asked himself, prompted by Loschmidt's complaints mentioned above)? This points to a serious conflict between our knowledge of the world and our experience of that world: our memory records (of, e.g. cups of tea that were *not* shattered on the floor) are more likely to be false than true, since the events they seem to concern are improbable (thanks to their lower entropy than the present condition).

Boltzmann indeed surmised that followed far enough into a distant past, we would find maximal entropy there too (molecular chaos), just as in the future. The asymmetry smears out over long enough temporal journeys. This is truly a mind-bending proposition, every bit as radical as the idea that the Earth is not at a distinguished point of the universe. It might seem that if we know why it goes up in the future (using Boltzmann's idea that the increase would be perfectly natural if the past contained a local fluctuation of very low entropy) then we have explained the arrow. But we are left with a puzzle about the past in this case.

One explanation, that we have already met in connection with time's arrow, is to simply *postulate* a low-entropy past (the 'past hypothesis': a label due to David Albert). This is usually attributed to special conditions of the Big Bang in our universe's past. Hence, the idea is that if our universe began in an *immensely* low-entropy condition, then it can't help but to increase in entropy, such as we find (as mentioned above).

To attempt to explain the increase of entropy by postulating a low-entropy past simply introduces another problem (or, in some ways, recapitulates the same problem): how do we then explain the low-entropy past? It is not enough to point to the Big Bang and utilize the apparent low entropy found there, since that very order is what is in need of explanation. Passing the buck onto something so incredibly improbable as the Big Bang leaves us no better off. However, the past hypothesis does have the considerable virtue of defusing the Loschmidt reversibility objection: our time asymmetry (linked to the entropy-asymmetry) rests on the existence of asymmetric boundary conditions. The unlikely configuration at the absolute zero of time as we know it (the Big Bang event) means that whatever patch of order we find, we know that preceding it will be a patch of lower entropy, tracing all the way back to the initial temporal boundary that the Big Bang provides (and at which we impose our boundary condition).

But even supposing we could resolve this issue, how plausible an account of time's arrow is it anyway? How plausible is it that all of those many irreversible features – including the asymmetries of our memory records, the non-unmixing of our tea, our unfortunate ageing – are due to something that happened around fourteen billion years ago?!

One way to think about this is to imagine what the world would look like if we had reached a state of maximal entropy, an equilibrium state in which there is one temperature for all. Would there be a direction of time in such a world? It's hard to see how asymmetric processes could be taking place. To use the standard metaphor of running the video of such a world backwards, we would find no distinction, and hence there would be no physically observable arrow of time. Though we could still speak of a second law in operation in such a universe, there would be nothing for it to 'act on' as it were. There would therefore be no temperature gradient, no entropy gradient. So we can see how having an entropy gradient *is* necessary in some sense to allow the sorts of time-invariant changes we experience, and if the gradient 'flows' to a state of maximum entropy, then we can see how the story is supposed to work. But a necessary condition is not quite the same thing as an explanation.

In one sense, it is good to ground *universal* behavior in a *common* origin. However, again, the way this is done in the present case simply pushes the mystery of time's arrow one step back. We immediately want to ask: why *those* initial conditions? Unfortunately, our present physical theories don't enable us to provide an answer – though plenty are trying!

There are two options: explain the present improbability, or don't. Boltzmann was in the latter camp, not thinking it plausible to deduce our improbable past and present from fundamentals. For Boltzmann the universe at a large scale is in thermal equilibrium (as the most probable of all states), and our existence is due to a local deviation (a quirk of probability) – see fig. 6.8.

To wax lyrical for a moment: we are born of random fluctuations, and will return to random fluctuations. Hence, he denied any fundamental status to the arrow of time, viewing it as a purely *local* phenomenon, much as the convention of 'up' and 'down' on the Earth is local (see fig. 6.9). The arrow of time stems from an Anthropic coincidence, that there happen to be sentient creatures (us) supported by the local fluctuation. Given our circumstances, of finding ourselves in an ordered world (how could it be otherwise if we are to exist?), we will see an arrow of time in which the improbable order moves to disorder. And the crucial point is *this applies to both directions of time if one looks far enough back* (i.e. to before the low-entropy blip). Boltzmann predicted, from his analysis, the existence of a kind of multiverse picture, in which scattered about the universe, which, remember, is in thermal equilibrium (essentially dead, with no elbow room to increase in entropy), there are worlds like ours (fluctuations about equilibrium), which depart from equilibrium for short times compared to the timescale of the entire universe (see fig. 6.8). Just as in modern multiverse theories one needs an enormous amount of worlds to explain the existence of our apparently fine-tuned universe, so for Boltzmann one needed an immensely vast universe in thermal equilibrium to explain our world – see §8.7 for more on Anthropic reasoning. Were we not seeing an arrow of time,

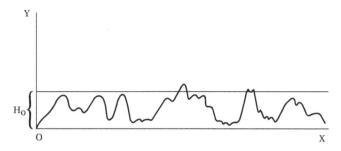

Fig. 6.8   Boltzmann's diagram showing the local deviations from thermal equilibrium. Reproduced in S. Brush, ed. *The Kind of Motion We Call Heat* (North Holland, 1986: p. 418). Here *H* is a quantity that Boltzmann himself used, inversely related to entropy: when it is low, entropy is high and vice versa. The curve represents the history of the universe, where the highest peaks (or "summits") correspond to phases potentially containing life (such as our own world): the larger the peaks, the more improbable the state. (Here *y* is the 'extension' of the state giving its probability of occurrence and *x* is the temporal direction.)

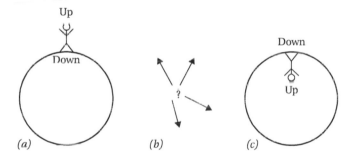

Fig. 6.9   'Up' and 'down' orientations are locally defined relative to reference frames. In (a) and (c) a person's conventions for orientation will be distinct depending on whether they live on the surface of a sphere (such as the Earth) or in the interior of a sphere (such as aboard a rotating spaceship). With no such reference systems to fix a convenient orientation convention (as in (b)), 'up' and 'down' are meaningless.

the entropy would be too high for an arrow to emerge (not enough gradient), and so we wouldn't be here to consider it. In other words: wherever there be observers, there be an arrow of time. It couldn't be otherwise. Moreover, the local conditions would dictate which direction would be considered 'future' and which 'past' (see fig. 6.10).

Other attempts at an explanation focus on aspects of early universe physics in an attempt to fill in the details of the 'special initial conditions.' One famous example of this, due to Roger Penrose, invokes the 'Weyl curvature hypothesis' – where the Weyl curvature tensor tracks aspects

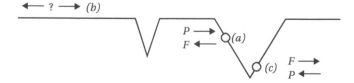

Fig. 6.10   By analogy with the 'up' versus 'down' scenario, here we see three possibilities with respect to time's arrow: in (*a*) the past runs toward the right (and the future to the left), since that represents a decrease in entropy (a larger deviation from the thermal equilibrium); in (*c*) the past runs toward the left (and the future to the right), since that represents a decrease in entropy relative to its location; in (*b*) there is no entropy gradient to determine an arrow of time.

of the curvature of spacetime that would be found in space emptied of all but gravitational field (matter is ignored here since it is believed to be at maximal equilibrium in the earliest phases). Penrose's hypothesis is that this entity vanishes near the Big Bang singularity and implies a low entropy there. But this won't do as an *explanation* of the past hypothesis (the special initial state); instead it *reformulates* the past hypothesis *in terms of* the Weyl tensor. Why do we have the vanishing Weyl tensor is simply a way of re-expressing the original puzzle. Indeed, Penrose passes the buck to a future theory of quantum gravity that would be able to deal with situations in which gravitational fields are incredibly large, and quantum effects must come into play.

We saw earlier that Zermelo was in the former camp (*explain* the improbable state) – or at least, he viewed the bruteness of the improbability as a *problem* for the statistical-mechanical approach. Boltzmann thought this was unprovable and had to be assumed.

In terms of explanation of the low entropy, one might extract an Anthropic explanation from Boltzmann's own arguments as above. After all, he believed in what is essentially a vast ensemble of worlds, and what kind of world would we find ourselves in but one with a low entropy past? Of course, as an explanation it is weak, since we aren't deriving our world from the scheme, only showing how they are compatible (which was all Boltzmann was concerned with). When we push Boltzmann's idea, however, we find it faces a curious problem (perhaps a *reductio ad absurdum*), which we turn to next. The problem, simply put, is that if it's consistency with observable results that we're after, then it can be achieved on a tighter entropy-budget than generating an entire universe that has persisted for billions of years: what a waste of effort! If one could get the same effect (a universe, that appears to have existed for billions of years, with star formation, the evolution of life, etc.) from a much smaller fluctuation – say a single Cartesian-style disembodied brain configured with the correct memories, believing it has had the 'correct' experiences, and so on – then, given the inverse relationship between the magnitude of the fluctuation away from thermal equilibrium and the probability of such a fluctuation occurring, this brain scenario would be *vastly* more likely. This is, of course (I hope!), absurd. If we are bothered by the problem of other minds (so that we cannot countenance solipsistic single-mind views of the universe), then there are still many more likely scenarios than having the universe actually be as old as it seems (which is a persistent fluctuation, and so less likely than a shorter-lived one): the theory that the universe came into existence five minutes ago (fully formed) is far more likely in terms of the likelihood of such a fluctuation relative to an extended one trailing billions of years into the past.

6.8 Typing Monkeys and Boltzmann Brains

Fans of *The Simpsons* might recall the episode 'Last Exit to Springfield,' in which a kidnapped Homer is shown a room in Burns' mansion in which a thousand monkeys are typing away, with one typing the almost correct "It was the best of times, it was blurst of times," from Charles Dickens' *A Tale of Two Cities*. The so-called 'monkey theorem' tells us that given enough patience (and *time* for that patience to manifest itself), and some monkeys randomly thwacking keys on a computer, we can expect to see generated the complete plays of William Shakespeare. One would, in fact, require a *lot* of time: far more than the present estimates for the age of the universe (about fourteen billion years). Amusingly, a computer programmer, Jesse Anderson, attempted to approximate this by simulating millions of monkeys ('virtual monkeys'), which churn out random ASCII characters between A to Z, which are then fed to a filter to test for fit with actual lines of Shakespeare (as found on Project Gutenberg) – the results are as expected: one can randomly generate such apparently meaningful segments given time enough and monkeys or monkeys enough and time (i.e. processing power capable of generating the appropriate output strings). Of course, each string is as random as the other (by construction), but we *see order* in certain strings. To get a handle for how approximate such a simulation would be, just consider that for a work with 500,000 characters (a reasonably sized novel), the probability of getting the novel churned out in this random way would be 2.6×10^{-500000} (3.7×10^{-500000} if we add numbers 0 to 9 and blank spaces). By the same token, one can see that if time were no obstacle, it *would happen* sooner or later – hence, this is often labeled the 'infinite monkey theorem': one monkey with infinite time, or infinitely many monkeys with finite time, either will do the job.

We can play a similar game (known as Boltzmann's brains) with atoms, time, and space, only this time constructing (now via fluctuations from the vacuum) any physical structure we could care to name, including ourselves and the entire observable universe. Indeed, given a universe with infinite time, you will be 'reincarnated' infinitely often: eternal recurrence! There are vastly more possible configurations than constrained by the choice of ASCII characters in this case, so the probabilities will be vastly smaller. But the point remains, given unlimited time, it would happen – the 'principle of plenitude' (that anything that *can* happen *will* happen) likes infinite time. (Interestingly, one can find a strikingly similar scenario painted in Friedrich Nietzsche's combinatorial notion of *eternal recurrence*. Simply expressed, he points out that given some definite constant energy distributed over a finite number of particles, in infinite time every combination must already have happened, and infinitely many times, repeating in cycles as each initial state recurs. In Nietzsche's mind this was supposed to be a victory against the materialistic physics that pointed to a winding down of the universe, in a final equilibrium state known as the 'heat death.' Yet what such recurrence objections reveal is that the second law does not apply universally, and that Boltzmann's statistical, yet still mechanical, viewpoint is to be preferred.)

The implications here are similar to the classic 'brain in a vat' scenarios that epistemologists like to chew on. From the perspective of us, 'inside our own skulls' as it were, we can't tell whether we have genuinely lived the lives we appear to have lived, having the various experiences that formed the memories we appear to have, caused by a universe that appears to be very old, and so on. We might be a disembodied brain in a laboratory undergoing stimulation and reconfiguration to generate such appearances. Our memories would not be records of real events; rather, they would be implants. As David Albert emphasizes, looking in history books for evidence is no good: they too are more likely to have fluctuated into existence without any causal link to the events (e.g. the rise and fall of the Roman Empire) they appear to describe – this is analogous to Dr Johnson kicking a rock to refute George Berkeley's brand of skepticism about the external world.

Likewise, in the case of Boltzmann's brains. In a universe that persists for long enough, there is a tiny but non-vanishing probability that a brain with all of your thoughts and memories, thinking it has lived a life having read books about physics and so on, will spontaneously generate. Given that the energy cost of generating such a region of order (the disembodied brain) is considerably smaller than generating an entire universe whose development over fourteen billion years led to it, it is *more likely* that we are such a Boltzmann brain than not! Boltzmann brains ought to be more *typical* than the long-haul brains we think we have. To put it another way: If you happened to be a god looking to create sentient beings, you would save a lot by just bypassing galactic and biological evolution, and having them pop into existence fully formed.

We have seen that Boltzmann's explanation for the thermodynamical arrow of time (entropic increase) says that regions of time asymmetry are unusual: the usual state is one of molecular chaos. Hence, we are in a special state: time-symmetry is the norm. From this thermal bath, all manner of odd fluctuations can arise, with the size of their deviations from equilibrium linked to their likelihood of occurrence. Sean Carroll puts the point rather nicely:

> [A] universe with a cosmological constant is like a box of gas (the size of the horizon) which lasts forever with a fixed temperature. Which means there are random fluctuations. If we wait long enough, some region of the universe will fluctuate into absolutely any configuration of matter compatible with the local laws of physics. Atoms, viruses, people, dragons, what have you. And, let's admit it, the very idea of orderly configurations of matter spontaneously fluctuating out of chaos sounds a bit loopy, as critics have noted. But everything I've just said is based on physics we think we understand: quantum field theory, general relativity, and the cosmological constant. This is the real world, baby. ("The Higgs Boson vs. Boltzmann Brains": http://www.preposterousuniverse.com/blog/2013/08/22/the-higgs-boson-vs-boltzmann-brains/)

If this is the implication of our best physics (as it seems to be), then I submit that it needs the urgent attention of philosophers (in addition to physicists)!

6.9 Why Don't We Know about the Future?

This might sound like a silly question (the future hasn't yet happened, stupid), but it clearly has to do with the arrow of time and, as we have seen, statistical mechanics and entropy are bound up with that. Perhaps, then, one can explain the 'epistemic arrow of time' (involving the past–future asymmetry of *knowledge*) by making use of some of these concepts? Since entropy increases from past to future, we need to make sense of how it can be that our present observations tell us more about the past than the future.

The classic example involves the notion of a *memory trace*. A trace is a low-entropy state. A memory record, for example, must be sufficiently distinctive to 'make a mark' on the host. Something that wasn't there previously. One can think of this process (and the link to an experience of a flow of time: a subjective feeling of a sequence of events seemingly *coming into being*) in terms of the difference between randomness and order. A uniform, random experience or event (say of a perfectly flat featureless sandy beach) will not be distinguishable in itself (assuming all other conditions are kept fixed) from a snapshot of the same beach at a later time. However, if a wind stirs up a sand dune (especially if that dune takes on the appearance of some object or person, say) then it leaves a mark. It is a departure from randomness.

This (marks, traces, etc.) is really all we have to go on evidentially speaking, and so we make inferences about the past (retrodictions) just as much as the future (predictions). But we seem to have a stronger basis for past inferences: though I might be mistaken about the exact way in which some past event occurred (inferring from a memory or some other trace), I know that *something* happened; yet with the future we don't seem to have such certainty – I might think 'next Monday there's a research seminar,' yet some calamity could cause the speaker to cancel so that it doesn't occur. So how does a notion of trace help here? Hans Reichenbach's case of a footprint found on a sandy shore is a good example of a kind of trace. One can also readily see the link to entropic considerations. The footprint (trace) provides present evidence about what happened in the past. The idea is that the footprint 'does not belong there,' entropically speaking, relative to its surroundings, enabling one to infer that there was some kind of intervention (an outside influence) leading to its existence. In other words, the sandy shore is not a closed system, and an entropy-reducing process caused the local reduction in entropy characterizing the footprint in the sand. The shore's entropy was lowered by something external: it can't have been a closed system.

But just how relevant is entropy to such retrodictions from traces? John Earman argues that it is not at all relevant: one can just as well make past inferences on the basis of very *high* entropy. Low entropy is not necessary for traces, and one can imagine such things as catastrophes wreaking havoc on some orderly town, increasing the entropy there. Yet one could still make an inference, from the damage, that there was an intervention 'from outside.' Given the nature of the damage caused, one can even describe features of the causal influence (e.g. using the Saffir-Simpson wind scale and other such instruments). However, as Barrett and Sober point out, though the low-to-high transition of entropy is not at work, one should not rule out the importance of entropy in inference. One can say that, given the system (town, beach, etc.), there are certain expected entropies that if disturbed (in either direction) point to outside influences. But in such cases, *background knowledge* of the system (providing history and standard behaviors) is needed. Hence, the crucial aspect is that there is some departure from *expected* properties (such as entropy).

Note also that the marks seemingly provide evidence for our inferences about the past only if we accept the past hypothesis. It is the past hypothesis that provides us with the explanatory edge that we seem to have with the past over the future. Otherwise, we might infer that the fluctuation from molecular chaos story was the correct one (since it would be the most statistically likely). Some inferences about the past and future are balanced:

those that come from the information about the current state combined with the laws, which allow us to propagate states in both directions. But we can do more into the past. Suppose, for example, that your mother walks in and sees the broken cup on the floor. She doesn't assume it's a random fluctuation, but that it was the result of an ordered causal sequence of progressively lower entropy states – unless she's a physicist, she's unlikely to think of it this way, but her thought-processes will be along just such lines all the same: 'the cup was together in one piece, and contained hot tea, which involved it being contained within a kettle . . .'. (You might compare this additional way of reasoning with Laplace's demon armed only with the laws and the initial state of all the matter in a system.)

Bizarrely then, the asymmetry of knowledge (and the function of memory) *depends* on the conditions many billions of years ago. If we didn't utilize it (however implicitly) our reasoning about the past would be no better than about the future. Our reasoning about the future is uncertain about many things, but we know that molecular chaos will be the likely event of most present (well-ordered) conditions. The past hypothesis means that the same is not true of the past (and so the conditions that led up to the well-ordered present event) You might, as mentioned earlier, find the linking of memories (which are also clouded with all sorts of emotions) with the low-entropy past as too flimsy to really provide an adequate explanation. Still, it would then remain as a challenge to explain how memories give us information about the past.

To finish up this fascinating issue, I leave the reader to ponder what other areas of our inferential machinery are dependent on the past hypothesis.

6.10 Further Readings

Some excellent recent textbooks on the philosophy of statistical mechanics have emerged recently. Also, thanks to the links to such issues as arrows of time and knowledge there are also some fun, popular books.

Fun

- David Albert (2001) *Time and Chance.* Harvard University Press.
  - Short and sweet! Essential reading for any philosopher of physics, and anyone wishing to understand the deeper implications of statistical mechanics.

- Sean Carroll (2010) *From Eternity to Here: The Quest for the Ultimate Theory of Time.* Dutton.

  - Popular, and with sparkling examples, but the issues are presented with real depth of understanding.

Serious

- Huw Price (1997) *Time's Arrow and Archimedes' Point: New Directions for the Physics of Time.* Oxford University Press.

  - Comprehensive treatment of time asymmetries covering some very difficult terrain in an easy to understand way.

- Lawrence Sklar (1995) *Physics and Chance: Philosophical Issues in the Foundations of Statistical Mechanics.* Cambridge University Press.

  - Solid text covering virtually all the central topics in philosophy of statistical mechanics.

Connoisseurs

- Meir Hemmo and Orly Shenker (2012) *The Road to Maxwell's Demon: Conceptual Foundations of Statistical Mechanics.* Cambridge University Press

  - The focus is on issues of time asymmetry, but this book covers a very wide range of issues in the foundations of statistical mechanics in great depth and in an exceptionally clear way.

- Gerhard Ernst and Andreas Hüttemann, eds. (2010) *Time, Chance, and Reduction: Philosophical Aspects of Statistical Mechanics.* Cambridge University Press.

  - State of the art treatment of time's arrow, the meaning of probabilities, and reduction.

7 Quantum Philosophy

Quantum mechanics is our best theory of the material world. It has been applied successfully to three of the four interactions of nature (electromagnetic, strong, and weak), and work continues apace to apply it to gravitation. Scott Aaronson ([1], p. 110) describes quantum mechanics as not so much a physical theory, but as something falling halfway between a physical theory and a piece of pure mathematics. He uses the analogy of an operating system [OS], where the procedure of making a theory quantum mechanical (i.e. quantizing) amounts to 'porting' the application (e.g. Maxwell's classical theory of electromagnetism) to the OS. One might extend his analogy by thinking of quantum mechanics itself as a significant 'upgrade' from classical (Newtonian) mechanics, which proved unable to 'run' certain programs. All of our quantum theories are achieved through this porting procedure: one starts off with a known classical theory and then performs some specific modifications to it.[1]

Porting an application into the quantum OS brings along with it a whole bunch of curious features that did not appear on the older OS: indeterminacy, matter waves, contextuality, non-individuality, decoherence, entanglement, and more. These features still account for the vast majority of work done within the philosophy of physics, though recent work done on 'quantum computation' has altered their flavour somewhat. Other recent work on quantum theory has tended to focus on specific issues of quantum *fields*, especially the issue of the extent to which the theory contains particles. These more advanced issues must wait until the next, final chapter. For now we focus on the 'classic' philosophical problems of quantum mechanics, and get to grips with its basic features. The four core problems we focus on are: the interpretation of probability and uncertainty; the measurement problem; the problem of nonlocality; and the problem of identity. These overlap and splinter in a great variety of ways, as we will see. Firstly, let us motivate some of the basic oddities of quantum mechanics.

7.1 Why is Quantum Mechanics Weird?

I'm sure that anyone reading this book will have heard all of the sayings about how strange quantum mechanics is. The quotes from famous physicists are legion: 'if you think you understand quantum mechanics, then you don't understand quantum mechanics.' In his popular book on quantum electrodynamics, Richard Feynman puts it like this:

> What I am going to tell you about is what we teach our physics students in the third or fourth year of graduate school – and you think I'm going to explain it to you so you can understand it? No, you're not going to be able to understand it. Why, then, am I going to bother you with all this? Why are you going to sit here all this time, when you won't be able to understand what I am going to say? It is my task to convince you not to turn away because you don't understand it. You see, my physics students don't understand it either. That is because I don't understand it. Nobody does. ([14], p. 9)

Often, amongst physicists of a certain stripe, thinking about the *meaning* of quantum mechanics is a violation of some unwritten rule of what physicists are supposed to do. Or worse, it might lead one into philosophy talk! The slogan is: 'shut up and calculate!'

But quantum mechanics is a *physical* theory. Experiments demonstrate quite clearly that it applies (even if it is ultimately only an approximation) to the world (the *actual* world): our world! Surely it ought to be understandable? We ought to be able to say something about *how* the theory latches (with such impressive empirical success) onto the systems in this world. That is, there ought to be some *interpretation* of the theoretical formalism that enables us to see how the theory 'works its magic.' There is no such thing as magic, so there must be some rational, physical account. And indeed there is, or rather *are*.

There exist very many ways to 'make sense' of quantum mechanics: Copenhagen, modal, relative-state, many-worlds, many-minds, Bayesian, Bohmian, Qbist, spontaneous collapse, etc. However, what one person considers to be a perfectly rational account, another might consider to be outright lunacy. There is an almost religious fervour concerning the holding of a particular stance on quantum mechanics: 'the church of Everett' versus 'the church of Bohm'! Let us make a start on finding out what it is people are disagreeing about – it sure isn't anything experimentally testable, which is why many physicists dismiss the whole business of interpretation as completely irrelevant. However, whatever position one adopts, it is undeniably true that there are elements of quantum mechanics that are genuinely weird, as we will see – but in many ways, no less weird than time dilating and spacetime warping in the context of the theories of relativity. Perhaps the weirdest is the quantum version of the two slit experiment.

As Feynman maintains in his *Lectures on Physics*, much of the strangeness of quantum mechanics can be seen in the so-called double slit experiment, which he argues is "impossible, *absolutely* impossible, to

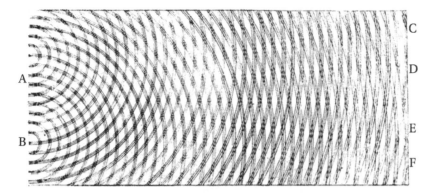

Fig. 7.1   The interference pattern as drawn by Thomas Young for a wave passing through a screen with two slits.

explain in any classical way" ([12], p. 1) – indeed, he believes this contains "the *only* mystery" in quantum mechanics, and one that cannot be gotten rid of by explaining how it works (ibid.): there is no explanation, only (extraordinarily precise, though still probabilistic) prediction. This experiment reveals the interference (wavelike) properties of quantum particles. It is this interference that's responsible for most of the other curiosities of quantum mechanics – including the speed gains that quantum computers make over classical ones. A similar experiment was used earlier, in 1802, by Thomas Young to demonstrate the wave nature of light (and thus disconfirm Newton's corpuscular theory) – see fig. 7.1. Quite simply, as the light travels from the two slits, A and B, to the detection screen it will have sometimes shorter and sometimes longer distances to reach the various parts of the screen. There will be cases where light following a short path from one slit coincides with light following a long path from the other slit, and so the light is likely to be out of phase at such points. One can then have either constructive or destructive interference, which will give rise to lighter bands (when the phases add) and darker bands (when the phases subtract) on the screen, at C, D, E, F.

It is one thing for a beam of light to behave in this watery way, but the startling feature of quantum mechanics, known as 'wave-particle duality,' is that it also applies to 'material particles' too. Inversely, thanks to the duality, particle-like behavior can be found in phenomena more commonly thought to be wavelike (such as light). With modern technology, one can perform these experiments so that only single particles are leaving the source and traveling to the screen, leaving a single click or scintillation where they are detected (see fig. 7.2).

After a small number of detection events the clicks appear to be random. However, on performing many such runs one finds the most startling result: even though the particles are hitting the screen as individual events,

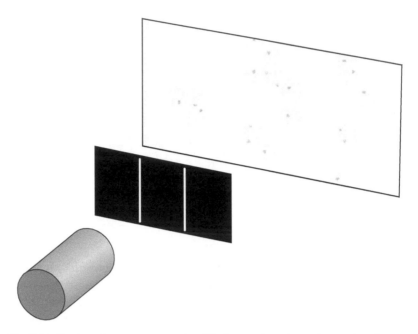

Fig. 7.2   The interference pattern after 65 photons have been detected at the screen. This simulation was carried out in *Mathematica* with a slit separation of 1 cm and a wavelength of light of 560 nm. [Code by S. M. Binder: http://demonstrations.wolfram.com/WaveParticleDualityInTheDoubleSlitExperiment/].

over time the old classical wave pattern (as sketched by Young) is built up much as a pointillist painter discretely renders a continuous scene (see fig. 7.3).

We have used light in this example, but as mentioned above, the same results can be found with electrons and other particles. Indeed, one can even find this kind of behavior for complex structures such as molecules (including organic molecules) consisting of almost 1,000 atoms. The difficulty in 'supersizing' the double slit experiment lies in upholding the 'quantumness' (quantum coherence) of the objects in the face of *decoherence*, which destroys the interesting phase (interference) effects through interacting with the environment and its many degrees of freedom.

To add to the oddness of this experiment, when one performs the experiment with just one slit open the strange wave-like pattern vanishes and is replaced by the boring classical buildup one would expect. Yet why does the particle care whether one or two slits are open: surely it goes through one or the other, not both? However, it seems to in some sense go through both – even a single particle – constructively and destructively interfering with itself, and then becoming a point-like particle once again when it reaches some measurement device (such as the detection screen). This apparent measurement-dependence of wave (continuous, linear) versus

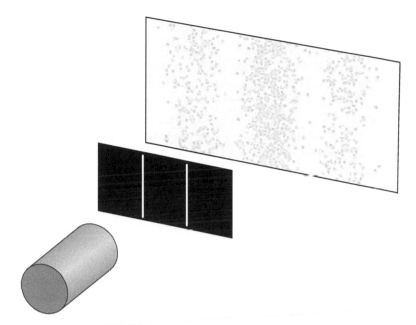

Fig. 7.3   The interference pattern after 3,500 photons have been detected at the screen (for the same slit separation and wavelength settings – note that altering these can alter the spacings between the bands). Is it a particle? Is it a wave? No, it's neither (or both)!

particle (discrete, non-linear) behavior is part of the quantum measurement problem, and it, rather than superposition itself, standardly supplies the proving-ground of interpretations of quantum mechanics.

Let us apply the language of wavefunctions to this setup. Remember from Chapter 2 that to each state of a system there is associated a wave-function ψ. From this we draw a probability P for being found to possess some particular property (e.g. to be at a particular location x on a screen). Let $\psi(x)$ be the probability amplitude (a complex number) for the probability (which as an amplitude can be either positive or negative) of a particle in the two slit setup to hit the screen at a distance x from the center of the screen (where the center is the point lying equidistantly between the two slits). $|\psi(x)|^2$ is then the *probability density* whose integral over some interval gives the probability for finding the particle in that interval. In the case of the two slit experiment we clearly have two options (two mutually exclusive routes): 'the particle goes through slit 1 to get to a point x on the screen,' or 'the particle goes through slit 2 to get to a point x on the screen.' Represent these two possibilities by the amplitudes $\psi_1(x)$ and $\psi_2(x)$ respectively (associated with probabilities P_1 and P_2 respectively). The probability density for the particle to make a click at x is then:

$$P_{12}(x) = |\psi_1(x) + \psi_2(x)|^2. \tag{7.1}$$

The density function will be peaked on $x = 0$ (directly between the two slits on the screen) and also on integer multiples of $x = \pm\lambda D/s$ (where s is the slit separation, D is the distance between the slits and the detection screen, and λ is the wavelength of the beam).

Feynman's remark about there being no classical explanation for this stems from the fact that a classical explanation would involve the simple *additivity* of probabilities of a particle going through slit 1 *and* a particle going through slit 2. In other words, for a classical particle theory we would have the sum:

$$P_{12}(x) = P_1 + P_2. \tag{7.2}$$

That is, there is no interference between the alternative possible outcomes (represented by $\psi_1(x)$ $\psi_2(x)$): we just have to add together the (classical) probabilities for the separate events, here P_1 and P_2, which would give a distribution peaked at $x = 0$ again, but decaying more or less uniformly as $|x| > 0$ (and we move away from the center) rather than displaying the peaks and troughs characteristic of wavelike phenomena and interference, as we see in Young's diagram. By contrast the non-additivity of events for quantum particles is precisely what one expects of a wave. We must 'supplement' the classical probabilities with additional interference terms:

$$P_{12}(x) = P_1(x) + P_2(x) + I_{12}(x) \tag{7.3}$$

As we saw in §5.3, there is a variety of options when it comes to interpreting probabilities, so there remains a question mark over how we should interpret these quantum probabilities: are they objective or subjective – i.e. about the state of the world or the state of our knowledge of the world? Are they about a whole bunch of similarly prepared events (relative frequencies) or are they about individual events (propensities or something of that sort)? Depending on how one thinks about these probabilities (objective versus subjective, or ontological versus epistemic to use an alternative terminology) one is faced with the view that quantum mechanics is either *complete* (so there is a fundamental limit to what we can know about the world: it is fundamentally probabilistic) or it is *incomplete* (so that there is perhaps some deeper theory that can explain and predict the funny behavior in the two slit scenario: there are 'hidden variables' that quantum mechanics misses). The same applies to the quantum states themselves, of course: since this is our central object of interpretation, either quantum theory is *about* world-stuff or knowledge-stuff.

So: the double slit experiment is closely related to one of the first conceptual questions to be asked about quantum mechanics: whether the wave-function gives a complete picture of reality or whether it is a step on the way to a deeper theory not subject to irreducible probabilities. If this is so, then what is the representation relation between ψ and the world? According to Einstein's 'ensemble interpretation' it doesn't in fact refer directly to the actual world at all, but rather to a non-existent distribution of many systems of the same kind. This is in order to make sense of the quantum statistics. We turn to Einstein's famous argument in which he attempts to establish the incompleteness of quantum mechanics in §7.3. First we turn to the nature of probability and uncertainty in quantum mechanics.

7.2 Uncertainty and Quantum Probability

We have already met probabilities in physics in the previous chapter. However, in that case they were understood in an epistemic sense: probability was simply ignorance of the *true facts*. In situations in which the "true facts" are hard to determine, because of the extreme complexity of the system, for example, a statistical approach is a natural step. But the usage of probability is a matter of convenience. If only we had enough computing power to track and predict the movements of the parts of a complex system, and the resolving power to figure out their instantaneous states, then we could, in principle, eliminate probabilities and speak purely in terms of certainties. Weather prediction, for example, is (as you well know) fraught with uncertainty. But we do not think of this uncertainty as a brute fact about the world. Rather, we think that we simply don't know (1) the initial conditions well enough to make certain inferences from them; (2) the laws well enough to feel confident about plugging in initial conditions (even if we did have them, since they are highly non-linear); and (3) we don't have computers capable of running the evolution to make precise (unique) predictions. Again, probability here reflects our ignorance, rather than the world's inherent indefiniteness. A 65 percent chance of rain tomorrow does not mean that the world is in a fuzzy state: we are in a fuzzy state!

In terms of the modeling of probabilities in such cases, we would think of them as measures over a state space in which those states are assumed to be uniquely mapped to definite physical states. The uncertainty is a measure of ignorance, rather than a measure of an objective feature of the world. In the case of quantum mechanics the 'uncertainty principle' is taken to express a more fundamental kind of uncertainty: there is a limit, integral to the laws of physics, according to which, for certain pairs of properties, we cannot know the values of both simultaneously with perfect precision.

A useful way of thinking about the uncertainty relations is in terms of the properties of the wavefunction as one switches between a well-localized position on the one hand and a definite momentum on the other. In the former case, the wavefunction is peaked at some point of space, with the momentum spread out. In the latter case, the wavefunction becomes a plane wave spread over the whole of space (in theory, out to infinity). Moreover, we can see that trying to localize a particle restricts its motion, which in quantum theory involves more energy: probing smaller scales demands greater energies, which is why ever larger particle accelerators are required to go to 'deeper' levels of reality.

However, the problem is that this switching is a function of whether a measurement is performed to determine the particular property. Born's statistical interpretation, designed to make some sense of this switching, contains this same intrusion of measurement. The squares of the amplitudes were taken to correspond to the probability of observing some value of the relevant physical quantity given some measurement designed to determine it. The apparently bizarre wave packet, spread out over space, is not 'really' a physical entity, but instead represents probabilities for localized, observable events. However, as we saw in our discussion of the double slit experiment, these probabilities are rather unusual in that they involve interference between the various localized alternatives.

Mathematically speaking, the uncertainty relations rest on the fact that position x and momentum p provide dual (physically equivalent though formally inequivalent) representations of a quantum system's state. They are related by a Fourier transform, which involves a reciprocal (inverse) relationship between the two representations: as the position amplitude $\hat{\psi}(x)$ is narrowed down, the momentum amplitude $\hat{\psi}(p)$ is spread (and vice versa). The uncertainty *principle* converts this mathematical result into a statement about our ability to gain information about a system's 'complementary' observables (i.e. observables standing in just such reciprocal relations: those that are canonically conjugate):

$$\Delta x \cdot \Delta p \geq \hbar/2 \tag{7.4}$$

Here Δx and Δp are simply the standard deviations (i.e. root mean squared) for particle position and momentum. In terms of joint knowledge then, if our grip on x is given by Δx, then our grip on Δp cannot be less uncertain than $h/2\ \Delta x$. In the extreme case $\Delta x = 0$, our knowledge of Δp is infinitesimally small (infinitely large uncertainty).

As mentioned, in quantum mechanics, if you want to know what is happening at smaller and smaller distances, then you have to increase the energy of the probe. Again, this is why the current generation of particle accelerators are so much larger than previous generations. Thinking in terms of using light to see smaller and smaller objects, this would simply mean that one needs shorter wavelength λ (higher frequency ν) light to uncover ever smaller objects. We can relate this wavelength to the momentum p as follows:

$$\lambda = \frac{\hbar}{p} \tag{7.5}$$

This inverse relationship (larger momentum implies shorter wavelength) parallels the uncertainty principle. To pin down an electron's position with light we would have to use a short-wavelength/high-momentum beam. Of course, wavelength determines a natural limit on what can be resolved: no features smaller than the wavelength can be discerned. The problem is, if we use high-momentum photons, as we must to find out the position with a high degree of accuracy, then we *kick* the electron with those photons, imparting an uncertainty in its momentum. But if we were to use photons with lower momenta, then one faces the problem that the position is not pinned down closely enough.

Thus, this reciprocal relationship between distance and energy scales parallels the famous uncertainty relations in quantum mechanics, and indeed Heisenberg attempted to prove the uncertainty relations using 'physical' arguments based on such reasoning – he envisaged a fictional microscope that fired gamma rays at the particle. Quite simply, to localize a particle requires bouncing something off it. Finer localization requires higher energies so that the bounce will be stronger. However, this bounce will cause the particle to be uncertain in its momentum values, essentially being nudged to new values from the imparted momentum.

According to Heisenberg, then, a physically motivated interpretation of the relations could be given, based on the notion of a *disturbance* effect caused by measurement interactions. In order to determine the value of some quantity, one has to *do* something to it, and this changes its state. For example, to determine the position of an object we might naturally try to localize it using sound, light, or some other form of radiation. Heisenberg himself originally used this argument to prop up an epistemic view of uncertainty: it is *our* limitations that forbid us from finding precise values of conjugate pairs, but this does not imply that such pairs do not have precise values. However, swept up in Bohr's interpretation, he succumbed to an ontological reading according to which there is no fact of the matter about the precise values: the values don't exist. Indeed, without some interactions between the system of interest and a measurement device, one must remain silent about that system. Hence, the uncertainty relations in Heisenberg's hands grew out of a wider philosophical stance concerning the meaning of physical statements: such statements must be associated with their means of being measured or bringing them about.

Einstein sought to produce counterexamples to this radical position (which he viewed as a denial of objectivity) by devising scenarios in which both quantities of a conjugate pair could have their values simultaneously pinned down in a way that defeated the uncertainty relations, much as Maxwell had attempted to beat the second law with his demon. However, Bohr is widely agreed to have gotten the upper hand. Another serious problem here is how to square the view that the wavefunction is a representation of knowledge with the very real phenomena that one finds in the slit and interferometry experiments.

Note that Bohmian mechanics faces no such issues over particle uncertainty since it essentially 'buys' definite positions for the particle using a nonlocal field, which 'guides' them (this approach is sometimes known as 'the pilot-wave' interpretation): hence, we have a hidden variables approach here, but a nonlocal one. Wave-particle duality is thus split apart into a particle, that is a particle at all times, and a wave, that is a wave at all times. Given this, the uncertainty relations are captured entirely by a 'disturbance view' rather than the uncertainty being a fundamental feature of the world.

There have been some recent experimental attacks on the measurement disturbance view of the uncertainty relations using 'weak measurement' (i.e. measurements that don't collapse the state onto its eigenvectors, so that an initial state is kept intact: an interference pattern would still be observed on a screen following a weak measurement, for example).[2] The idea is to measure some individual system state (such as the polarization of a single photon) but also measure how much the measurement disturbs the state. By doing a weak measurement prior to another measurement, the effect of measurement's kick can be determined (by making a third *strong* measurement of the first property weakly measured). The results in the experiments conducted so far have found less impact than Heisenberg predicted: measurement disturbance, according to these results, does not add much to the inbuilt uncertainty of quantum mechanics. Whether loopholes can be found or not (e.g. in the assumptions of weak measurements, and whether they in fact constitute measurement in a strong enough sense), it does show that what seemed to be a purely interpretative distinction (the extent to which uncertainty is an artefact of measurement or a feature of the world) can, in principle, be linked to experiment, thanks to a new tool.

7.3 EPR, Odd Socks, and No-Go Theorems

In terms of philosophy of physics, 1935 was a good year: the famous EPR experiment (where E = Einstein, P = Podolsky, and R = Rosen) was presented, along with Bohr's rebuttal, and also Schrödinger's discussion, which involved the first proper discussion of quantum entanglement. EPR was Einstein's grand challenge to Bohr's Copenhagen ('tranquilizing') philosophy. The core of the debate was over whether quantum mechanics could be viewed as a complete theory (Bohr's view), or whether it was fundamentally incomplete (Einstein's view).

At the root of the historical argument between Bohr (and others) and Einstein was the issue of *realism* (or, better, 'objectivity'): do the properties of objects have values independently of our observing them? In the early days, as seen in the previous section, it was supposed that there was a 'disturbance' triggered by a measurement that 'brought the value into being.' In other words, the very act of observation was something *special*. That is, observation *generates* or *produces* the values we observe in measurements: those values weren't realized in the system (were 'indefinite') before the measurement (unless the system had been expressly *prepared* in such a value by prior measurement). It was this feature that so disturbed Einstein – and not the statistical nature of the theory, as is often supposed (encapsulated in his 'God does not play dice' remark). One might quite reasonably side with Einstein on this: it does seem rather strange to think that the world does not have its properties given independently of our interactions (that 'the Moon is not there when nobody looks,' as he once put it to Bohr).

What also concerned Einstein was the apparent 'spookiness' of the interactions that would happen in distantly separated regions of space if quantum mechanics was supposed to provide a complete representation of reality. Locality was paramount: after all, it had been the central principle of both of his relativity theories in that action at a distance was explicitly ruled out. The notion of an objective reality is linked to this: the values an object has should not be linked to values had by another causally isolated system.

A real state (commonly called 'an ontic state' in modern discussions) is simply one that exists independently of our measuring it or knowing about it (an 'element of reality' in Einstein's terminology): it is *objectively* given in the sense that there is a fact of the matter about which values are exemplified at all times. If we measure such a state then we will be finding out *what its state was* (assuming there is no disturbance) beforehand. The states of a classical mechanical system can clearly be thought of in this way, as *revealed* by measurement. For example, in a Hamiltonian setting (the classical mechanics in phase space, from §2.2) our (instantaneous) states will be phase points $x = (q(t), p(t))$, which will determine unique trajectories. Any intrusion of probability here is associated with ignorance, in principle eradicable by supplying the additional information: let's do the same with quantum probabilities, says Einstein. Assume that the states of quantum mechanics could be filled in by hidden variables.

The so-called Kochen–Specker theorem already causes serious problems for the idea that quantum objects possess their properties in a simple 'common-sense' way, as suggested by such Einstein-style incompleteness claims. This conception of quantum properties is known as 'value definiteness': observables on quantum mechanical systems have definite values at all times, not just when measured. However, given an assumption of non-contextuality (that properties are possessed independently of which measurement we decide to perform), the Kochen–Specker theorem shows that the job of definite value assignments to all properties simply cannot be carried out. This is essentially a formal result having to do with the way observables are represented by Hilbert space operators in quantum mechanics. However, let's put the Kochen–Specker theorem aside and focus on the famous 'EPR experiment,' since this leads us to matters of entanglement and nonlocality.

Einstein famously defended 'local realism' (against quantum mechanics) in his debate with Bohr. This consists of two distinct components:

1   Physical systems have properties independently of our observing them [they have *objective* reality]
2   Physical systems that are 'spacelike separated' (with one lying outside of the other's light cone) cannot causally influence one another so that measurements on one affect measurements on the other [= principle of *locality*]

EPR tended to focus on position and momentum values, given the historical context fixed by Heisenberg's uncertainty relations. But let's see how their argument is supposed to work by looking at David Bohm's simpler version involving spin measurements (known as an EPRB experiment). Here we have a central source that generates a pair of particles, A and B, in what is called a spin-0 (singlet) state (where $|\uparrow\rangle_A \otimes |\downarrow\rangle_B$ means that particle A is spin up and particle B is spin down, so that the total spin cancels to zero: the symbol $\otimes$ refers to the tensor product used to combine several subsystems in quantum mechanics). Formally, we can write the resulting state for the experiment as follows:

$$|\Psi\rangle_{AB} = \frac{1}{\sqrt{2}} \left[(|\uparrow\rangle_A \otimes |\downarrow\rangle_B) - (|\downarrow\rangle_A \otimes |\uparrow\rangle_B) \right] \qquad (7.6)$$

This is an *entangled* state: we can't express the state in terms of separate, well-localized $|\downarrow\rangle_B$ and $|\uparrow\rangle_A$ states at the detectors, since the singlet state doesn't 'factorize' in the sense that the total wavefunction (representing the state of the system as a whole) cannot be expressed as a combination (tensor product) of the states of the parts. The particles are sent out to the left and right, to have their spin-components measured at detectors that lie at spacelike separation to one another (see fig. 7.4) – this is, of course, to implement the locality condition. The measured spins will always be perfectly anti-correlated because of the conservation of angular momentum: if the left particle is measured to have $+\hbar/2$ then the right will always be measured to have $-\hbar/2$ (and vice versa). Indeed, the *choice* of which direction to orient the detector is made once the particles are sufficiently separated to let the locality principle really get a foothold – since these orientations are chosen 'at a whim' (a 'free will' assumption), the particles cannot have conspired to set their values so as to establish their perfect correlations.

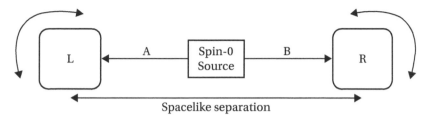

Fig. 7.4   A pair of particles in an 'EPR state' (a spin-0, singlet) is sent out to spacelike separation to be measured by a pair of reorientable spin-measurement devices (for the *z* component of spin). The direction of spin measured is chosen at each detector only once the particles are separated sufficiently to avoid any 'conspiring' to generate correlated results through direct communication (i.e. by local action or signaling).

The idea of EPR is then to compare 'common-sense' (as given by the locality and realism assumptions above, leading us to think that any measured correlations must be due to states had *before* measurements were made) with quantum mechanical predictions concerning the correlations we would find if we generated data on many such spin measurements. EPR claim that quantum mechanics fails the common-sense test by requiring nonlocal action (entanglement across spacelike separated distances). The common-sense view would simply point to the common origin in which the spin was zero, and so must be conserved. There would be perfect anti-correlation even when a free choice is made to alter the direction of the detector, rotating it by some amount – the spin-0 state has rotational symmetry, so it is impervious to such changes. A measurement will 'project' the state $|\psi_{AB}\rangle$ onto either $|\uparrow\rangle_A \otimes |\downarrow\rangle_B)$ or $- |\downarrow\rangle_A \otimes |\uparrow\rangle_B$, depending on whether particle *A* is measured to be spin up or down respectively. The problem raised by EPR is: how on Earth do the particles know what the other particle will 'reveal' on measurement so that it can coordinate itself accordingly? We can't run the classical 'common source' scenario in this case because that requires that the state would have been in an 'eigenstate' (think of this as a definite value for the measured property) all along from the source to the measurement so that it determines the measured result: but our singlet state is not of this kind. Worse, we have the freedom to choose our measurement direction (e.g. *x*-component instead of *z*) in mid-flight, while still preserving the perfect anti-correlated results – classically this would require that the information is built-in to the particles at the birth for *all* possible spin-measurements (including those that don't commute and so face the Heisenberg uncertainty relations). Again: how do the particles know what to do?

According to EPR there are just two possible explanations: (1) superluminal messaging allowing measured states to be communicated *at an instant* between particles, or (2) there is something missing from the quantum description of state, and this extra something is what determines the (anti-)correlations. Since the first option involves spacelike separated events, however, it seems that the measurement and subsegment distant effect could be switched by choosing an appropriate frame of reference, so that the link is not a Lorentz invariant notion.

As John Bell showed many years later (and as we see in a moment), it is possible to generate an experimentally testable difference between the common-sense (hidden variables) and quantum explanations: quantum mechanics will make different predictions to such a locally realistic theory. Let us just step back for a moment to consider an example of Bell's that makes the common-sense (classical) idea seem especially appealing.

The Irish physicist John Bell is often looked upon as an oracle by philosophers of physics, and not without justification: he is responsible for transforming the foundations of physics in such a way that philosophers of physics are likely never to be short of tasks. It is widely acknowledged that Bell's theorem, from his paper on the EPR paradox, was a much needed shot in the arm for foundational research on physics. It has been labeled 'experimental metaphysics' since it seems to rule out a metaphysical stance (Einstein's notion of 'local realism') – it was Alain Aspect who first realized that Bell's thought experiment could be made flesh, devoting his PhD thesis to the subject (though he performed it with photons and polarizers, with calcium atoms as the source).

Bell distinguished between observables, which we have met, and 'beables' (that is be-ables). The latter are supposed to be distinct from matters of observation and measurement: a system's beables constitute the values that it *has* rather than what it *will have* when it is observed. The old orthodox interpretation of quantum mechanics had it that the theory was all about what *would be* observed upon measurement, not about what *was* before measurement. In other words, measurements are not taken to reveal the pre-existing values of the measured particles, but in some curious way they 'bring about' such values. The famous EPR argument rests on just this distinction, with EPR taking the view that a sensible theory must be about beables, and Bohr (and followers) arguing that quantum theory, sensible or not, is about observables: things whose *raison d'etre* is to be measured.

In the case of EPR correlations, as we have seen, a natural response is that they are no more surprising than everyday correlations in which there is a past preparation making it the case that if one outcome is observed at one end of the experiment, another known outcome *must* be observed at the other regardless of the spatial distance that separates them at the time of measurement of either. Bell makes this intuition very clear with his story of Reinhold Bertlmann and his eccentric practice of always wearing odd socks:

> Dr. Bertlmann likes to wear two socks of different colors. Which color he will have on a given foot on a given day is quite unpredictable. But when you see that the first sock is pink you can be already sure that the second sock will not be pink. Observation of the first, and experience of Bertlmann, gives immediate information about the second. There is no accounting for tastes, but apart from that there is no mystery here.

We can even suppose that Bertlmann bundles together socks of distinct colors in his sock drawer at home, randomly grabbing a pair each day. The analogy with quantum particles and their properties looks fairly direct. In this case we suppose that a pair of particles is prepared in a singlet spin state, in which they are described by a single wavefunction in which the particles' spin-values are opposed to one another: if one is definitely spin-up the other is definitely spin-down. They are sent apart to enter a pair of widely separated Stern–Gerlach experiments (used to determine their spins along some given axis) whose magnets will either result either in the particle's going upwards or downwards (the Bohm version of EPR just mentioned).[3] Which will happen for any given individual experiment is only known with a certain probability, but one can say with certainty that if one value is found at one experiment then the opposite will be found at the other. One could, if so inclined, describe Bertlmann's socks by a wavefunction, and even speak of it as 'collapsing' when we notice the color of one of his socks (so that the composite state featuring both socks is fully known).

Of course, in the case of Bertlmann's socks one does not say that observing a pink sock on one of his feet causes the other sock to dramatically alter its ontological status to non-pink. We do not collapse his sock from a fuzzy to a definite state. The only thing that was fuzzy was our knowledge. Any assignment of uncertainty about sock color (represented in the wavefunction) is entirely epistemic.

The question is: what of the situation with quantum particles? Can we adopt this same epistemic strategy with them, so that the experiments are simply detecting (or revealing) the properties of the particles that they had all along? After all, isn't that what experiments are for: finding out what value some system had?

This is where Bell's famous no-go theorem enters.[4] It provides a *criterion* for deciding whether the correlations in your theory are like Bertlmann's socks or not. It also provides a route for testing whether our world is a Bertlmann-world or something more puzzling. Or, to put it Bell's way, we need to find a way of deciding whether quantum mechanics is local or nonlocal, and then we can figure out whether the world itself is home to nonlocal influences or not.

Firstly, we need to get a clearer grip on the central terms of the debate.

- **Einstein Locality:** this applies to multi-part systems in which the system's parts are spacelike separated. Then for some joint operator that is built as a product of the parts' individual operators (a superposition of the separate parts' states), its value will be built from the individual parts' values in the same way. The values of the individual components are independent from one another in the sense that a measurement of one does not interfere with the other.

The notion of a 'correlation' too should perhaps be spelled out. We know from the news that there are often stories pointing out a newly discovered link between some substance and a health condition. These provide us with various pairings: 'smoking and lung cancer;' 'coffee and Parkinson's disease;' 'marijuana and schizophrenia;' and so on. These are correlations rather than *causation* because the exact mechanism is not known: it is a *statistical* link. One might have found from some study that a large proportion of people who smoked a certain amount of marijuana also developed schizophrenia – more so than the general 'background rate' of schizophrenia in the population. So we could then assert that the probability of having schizophrenia *given* that you smoke marijuana is greater than if you didn't. But, so the saying goes, correlation does not equal causation. For example, it is possible that people who develop schizophrenia are more likely to self-medicate to alleviate the anxiety of stigmatization, so that the direction of influence is reversed (schizophrenia causes smoking, rather than the other way around).

A correlation once discovered is often a first step in filling in a deeper causal story, or in showing how some statistical error is confounding the true results. Hence, there is a task of explaining or explaining away a correlation once one has been found to occur in nature. There might be a variety of things leading to the presence of a correlation. There might be genuine causation going on, replete with a mechanism linking the two variables. For example, one might discover a gene that some people have that is 'switched on' by the some specific component contained in marijuana smoke. Unless one does lots of studies to reveal the generality of the correlation it might have been a simple coincidence that in this population studied there happened to have been more schizophrenics than would ordinarily be expected. I already mentioned above the idea that the schizophrenia might be leading to smoking as a form of self-medication. In this case we could seek some deeper brain disorder that would be responsible for both variables, thus providing a *common cause*. (The standard example of the notion of a common cause is in explaining the correlation between yellowing of the fingers and lung cancer. There is such a correlation, but neither causes the other: smoking cigarettes causes both. Philosophers speak of the common cause as 'screening off' the original correlation: smoking causally screens yellow fingers from lung cancer.) What we had initially would then be a 'causally spurious correlation.'

A final option is that the correlation is simply 'brute': an inexplicable feature of the world that cannot be further analyzed. This is, scientifically speaking, not the kind of thing we would wish for. However, as we will see, the correlations of certain quantum mechanical experiments are seen to be just like this according to many interpretations. The word 'correlations' should immediately indicate that we are dealing with statistics here: results of many experimental runs.

The correlations here are remarkably simple: if a particle is found to be spin-up on one side, then it will be found to be spin-down on the other side *regardless of how we establish the magnets' orientations* (so long as these orientations are set the same on both sides, anywhere between 0 and 360 degrees, the θ value: generally chosen to be a multiple of 120 degrees, so that there are three possible settings).

As mentioned already, a natural (common sense = Bertlmann's socks) response is that the particles were forged at the same source, and 'simply carry their instructions around with them.' These instructions (hidden variables) reveal themselves when measured and are responsible for what is measured. If we performed lots of experiments to determine this, we would find a distinctive set of results appearing: if there were no correlations, as we would expect given the large spatial separation forbidding direct causal interaction, then we would find that the joint probability distribution would factorize into a pair of individual probabilities for the experimental variables.

Take some general pair of outcomes A and B (which might be our spin up and down results), and take some adjustable setting values, a and b, that will determine what measurement is carried out on A and B respectively. Our concern is with the *joint* conditional probability distribution: $P(A, B|a, b)$ – the probability of getting outcomes A and B given (i.e. conditional upon) the settings a and b. If this does not factorize into independent probabilities, then we have a correlation: $P(A, B|a, b) \neq P_1(A|a)P_2(B|b)$. Common sense tells us to look for some causal reason for correlations. Think of yellowing fingers and lung cancer (a standard example). We find that $P(A, B) \neq P_1(A)P_2(B)$. Why? It doesn't seem that one can directly cause the other: no mechanism seems to link them. The trick is to add additional causal factors, a and b, so that we condition on whether those with lung cancer and those with yellowing fingers are also smokers. There might well be a few other factors, λ, such as genetic disposition to smoke, that further 'causally unlink' the yellowing and the lung cancer, tracing both back to some common origins. When we take these into consideration, we will find a joint probability distribution that factorizes:

$$P(A, B|a, b, \lambda) = P_1(A|a, \lambda)P_2(B|b, \lambda) \tag{7.7}$$

The 'settings' a and b are assumed to be independent: Bob's smoking in Kansas is not a likely cause of Vic's smoking in Kazakstan. So the EPR/hidden-variables idea goes. The λ in this case will be some, as yet undetermined, causal factors (the hidden variables) that when supplemented into our model of the EPRB experiment will explain the correlations without utilizing any spooky influences – they can potentially take many forms. The a and b are the freely chosen settings for the detector orientation; as with the Vic and Bob's smoking, these must be assumed independent (known as 'parameter' or 'setting independence') lest action-at-a-distance type influences enter the picture. This is in stark violation of special relativity. The similar independence of the outcomes (outcome independence), however, is not thought to be in violation of relativity. The difference is in the ability to *control* the parameters a and b (e.g. to signal), but not in the case of A and B (note that I am speaking of the outcomes and the experimental wings by A and B here). To sum up: a hidden variables model will say that the outcomes A must only depend on a and λ, not on the goings on with the b-settings. Fiddling with b should alter nothing about A, and at A we have no knowledge of these fiddlings due to parameter independence (implementing locality). Bell's theorem tells us that one of these independence conditions must be false.

We know also that if we consider $-z$ (or a 180-degree) rotation of the detector, then the two outcomes will then be perfectly correlated: anti-correlated for 0; correlated for 180. That leaves a whole lot of orientation settings that can be utilized independently at the two sides of the experiment. Hence, if we only focus on measurements made with the same a and b settings then we expect the outcomes to be (anti-)correlated, and there is no scope for distinguishing quantum and hidden variable models: we know that the probability of getting the same outcome is zero since we have a singlet state with opposite spins demanded. The interesting divergences come from considering general settings. Indeed, it is found that divergences in predictions occur when the relative angle between the settings differ from 0 and 180 (and also 90 and 270). The respective predicted correlations for hidden variables and quantum models are shown in fig. 7.5.

These correlations can be used to generate inequalities for the degree of correlation such that if the world has hidden variables, then the degree of correlation will be greater than or equal to some value, for example.

This graph (in fig. 7.5) also reveals that the predictions of a quantum model cannot be replicated by a (local) hidden variables model. The incompatibility provides the basis for an experimental test: each model will make a set of distinct predictions that will enable us to test whether our own world is quantum (non-local) or not. Quantum models will violate a feature (Bell's inequality) that a local, classical model will satisfy.

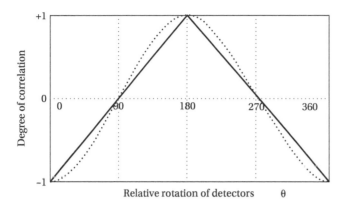

Fig. 7.5   The degree of correlation predicted by a hidden variables model (solid line) and a quantum mechanical model (dotted line) for various relatively rotated detector settings. Though there is agreement at multiples of 90 degrees, there is divergence elsewhere.

Experiments performed by Alain Aspect and others have since confirmed quantum mechanics over Einstein's hidden variables view of how the world should be. This leaves either or both of the two components of local realism at odds with the world. Either could in principle be denied. In the case of objectivity, one would have to give up on the idea that physics is about 'the way the world really is' (Bohr's view). In the latter case, one must accept the existence of superluminal connections: nonlocality. However, nonlocality in this case is rather different from nonlocal *signaling*, and the concept must be treated with care. Certain prohibitions elsewhere in quantum mechanics forbid the use of nonlocality as a means of superluminal communication.

There is some controversy over the validity of the results, and these work by finding flaws in some assumption of Bell's proof. One possibility, considered by Bell himself, is that the free choice of the experimenter is an illusion, so that the universe determines which settings will be selected. The universe thus orchestrates the correlations by determining what the experimenters decide to do (superdeterminism!). Here one must weigh up the relative implausibilities of nonlocality versus superdeterminism...

There is a related philosophical puzzle of how we are to *interpret* entangled systems ontologically speaking: what kind of world is a world containing entanglement. Lenny Susskind calls entanglement "the essential fact of quantum mechanics" (see p.xi of his book listed in Further Readings): it's what separates quantum from classical phenomena. Schrödinger, who coined the term in a 1935 paper, agreed:

> When two systems, of which we know the states by their respective representatives, enter into temporary physical interaction due to known forces

between them, and when after a time of mutual influence the systems separate again, then they can no longer be described in the same way as before, viz. by endowing each of them with a representative of its own. I would not call that *one* but rather *the* characteristic trait of quantum mechanics, the one that enforces its entire departure from classical lines of thought. By the interaction the two representatives [the quantum states] have become entangled. ([44], p. 555)

Entangled states, remember, refer to many particles (a composite system) that are in an eigenstate of some observable in a joint sense (as a pair), but not individually. For example, in the case of a spin observable, we might know that the total, joint system has spin 0, but not be any the wiser about the spins of the particles themselves, which can be measured up or down (+ or $-\hbar/2$). However, as we have already seen, we can say that if a measurement yields spin up on one particle (for some measurement orientation), then the other particle is spin down (for the same measurement orientation). The non-factorizability appears to lead to a kind of 'holism' since ontological independence of the systems can't be established – though a measurement will 'disentangle' them according to many interpretations. Here's Schrödinger again:

> Another way of expressing the peculiar situation is: the best possible knowledge of a *whole* does not necessarily include the best possible knowledge of all its *parts*, even though they may be entirely separate and therefore virtually capable of being the 'best possibly known,' i.e. of possessing, each of them, a representative of its own. The lack of knowledge is by no means due to the interaction being insufficiently known – at least not in the way that it could possibly be known more completely – it is due to the interaction itself. Attention has recently been called to the obvious but very disconcerting fact that even though we restrict the disentangling measurements to *one* system, the representative obtained for the *other* system is by no means independent of the particular choice of observations which we select for that purpose and which by the way are *entirely* arbitrary. It is rather discomforting that the theory should allow a system to be steered or piloted into one or the other type of state at the experimenter's mercy in spite of his having no access to it. ([44], p. 555)

In other words, we are free to choose, arbitrarily, which measurement to perform on one of the separated particles, but the disentangling will still occur for both particles regardless of their spatial separation. One spanner in the works for large spatial separations is that special relativity becomes relevant: this impacts on the total wavefunction for the system, which becomes non-factorizable (thus deepening the appearance of holism). However, at the same time the macroscopic distances involved in situations in which relativistic effects play a role make it hard to retain the coherence of the quantum state.

Though we see Schrödinger grappling with entanglement, and some of its odder implications (such as quantum steering and holism), it was left largely untouched until John Bell's work on the subject three decades later. Schrödinger himself thought that the nonlocality was unphysical. The past two decades or so have seen a radical transformation in the way entanglement is understood. Once viewed as a mysterious 'spooky' phenomenon, it now forms the basis of experiments and quantum mechanics. In more recent times, entanglement is so commonplace that computer scientists speak of it as a 'resource' to be manipulated and optimized.

Given the existence of genuinely entangled states, an interpretive task emerges for philosophers of physics: what kinds of things are they? This has proven to be highly controversial territory, with some arguing that entangled states call for entirely new ways of thinking about the world's ontology. Formally, remember, an entangled state (for an N-dimensional system) is 'non-factorizable' in the sense that a joint wavefunction for many particles cannot be written as separate wavefunctions for the subsystems: $\Psi_{1,...,N} \neq \Psi_1 \otimes \cdots \otimes \Psi_N$. Hence, there is a suggestion even in standard quantum mechanics that the world might not submit to being carved up in terms of 'individual things.' Entanglement involves the notion that the total quantum system is *not reducible to the intrinsic properties of its subsystems*, thus exhibiting a kind of 'holism.' Paul Teller developed a position known as 'relational holism' to capture this 'whole is more than the sum of its parts' aspect of quantum mechanics: the relations cannot be reduced to the non-relational properties of the relata (that is, they do not supervene on the non-relational properties on the relata). This aspect is known as 'non-separability': the inability to view the state of the world as reducible to local matters of fact.[5] This amounts to a kind of emergence too, since the emergent properties are not definable in terms of the non-relational properties of the subsystem parts. Likewise, Michael Esfeld and Vincent Lam [11] have argued that such relations in entangled systems can be understood as simply unreducible to the intrinsic subsystem properties. Ultimately, they propose that the 'object-property' distinction (where 'property' includes relations and structure) is not a fundamental ontological one, but a conceptual one. Hence, the phenomenon of entanglement links directly to deep matters of ontology and metaphysics.

7.4 The Quantum Mechanics of Cats

In a very tiny nutshell, the quantum measurement problem can be summed up as: quantum mechanics seems to allow superpositions to be amplified up to the macroscopic level. We don't see such superpositions of macroscopic alternatives. Therefore, we either need to find a way of burying them where they cannot be experienced or explain why they are not experienced in some way consistent with quantum theory (and thereby *solve* the measurement problem), or else reject the theory.

One might easily be forgiven for thinking, on the basis of much of the literature, that the quantum measurement problem is the only philosophical problem facing quantum mechanics. Its tendrils spread into many other areas in the conceptual foundations of quantum mechanics. The measurement problem follows quite simply from apparent facts of experience (the 'definiteness' of measured or observed properties) together with the assumptions of:

Linearity: time evolution under the Schrödinger equation is linear so that superpositions can 'spread' through interactions, so that if one system is in a superposition of its possible states, any system it interacts with will also evolve into a superposition of states;

Universality: the wavefunction ψ is able to describe *any and all* dynamical systems in nature.

With these two properties given we can infer that macroscopic objects will, like their microscopic cousins, be governed by quantum mechanics. This means that they too must have a linear state space, and so behave in a wave-like manner. This, in turn, means that we can have states such as (here we go!) a cat that is a linear superposition of both alive and dead (or neither alive nor dead) states. Such (macroscopic) states are, of course, never observed in nature. Rather, we observe *outcomes*: specific, unique, individual events: one or the other. Hence, we have a dilemma: (1) either our ideas about macroscopic objects are wrong; or (2) QM is false; breaks down at the level of macroscopic objects (such as a system capable of observing); or needs modifying in some way. This has some resemblance between the problems faced by the statistical view of entropy, with the conflict between theory and observation there: theory predicted a high entropy to the past as well as the future, but we observe this feature only in one direction. Similar mind-stretching (to Boltzmann's cosmological/Anthropic ideas) results in the context of the measurement problem in the attempt to resolve it.

This duality (wave behavior versus particulate behavior) is at the root of the measurement problem. The wave behavior is described by Schrödinger's equation. This is perfectly adequate for situations in which the system is not being observed or measured. However, seemingly, when it is measured a different kind of evolution occurs: a *disruption* of the linear, wave behavior. To give a simple example: suppose that we have some radioactive element that has a half-life of ten minutes. If we have a bunch of these atoms then after ten minutes have elapsed we can expect half to have decayed (into various decay products: radiation). If we consider a single such atom then after ten minutes *if we don't measure it* (e.g. with a geiger counter), then the quantum mechanical behavior will see it evolve into a linear superposition of decayed and undecayed: 50/50. The geiger counter, however, will *register* a definite event: click or no click. This transition from superposition to individual event constitutes a second form of evolution in quantum dynamics – though not all agree that it is a genuine feature of reality. This 50/50, 'blended' behavior so far is applied to the world of the very small, and so, though conceptually problematic, poses no direct observational problems – note that the 50/50 distribution is not necessary for the argument, which requires that a superposition be formed in which it is not the case that one state has probability 1 while the other states have probability 0: this latter is the characteristic of a measured state (which is another way of stating the problem).

The infamous Schrödinger's cat example makes use of this simple split in the kinds of evolution. The radioactive substance is coupled to some poisonous gas, both of which are within a closed box, such that if the substance decays then an unfortunate cat (also confined to the box) will perish. If it does not decay, then the cat is safe. However, we have seen that if a quantum system is not measured, it will enter a superposition of states, here: |decayed⟩ + |not decayed⟩. But the fact that the behavior of the atom is coupled to the poison implies that there will be an associated superposition of poison (prussic acid in Schrödinger's original example!) | emitted⟩ + |not emitted⟩. And, finally, since the cat's continued existence depends on the poison, and in turn the atom, it too will enter a superposition of |perish⟩ + |safe⟩. In other words, a superposition at the microscopic level (not directly observable) is amplified (through the evolution of the basic dynamical law of quantum mechanics) into a macroscopic superposition (with no definite component realized).

This obviously contrasts with the fact that we know our *experience* when opening the box to see what happened will reveal an individual event (one or the other component of the superposition). There seems to be, then, a conflict between quantum mechanics and experience! Macroscopic superpositions are predicted by quantum mechanics; but our experience reveals no such thing. This is the quantum measurement paradox. It is also another kind of incompleteness result since quantum mechanics seems not to be able to tell us about the state in the box.

What is bizarre about this is that it is most often assumed that our experience (of macroscopic definiteness) is veridical (true to the facts of the world) so that if quantum mechanics is also true, then we are somehow affecting quantum states when we perform measurements. This is known as *reduction* of the wave packet by observation or measurement, or more formally as 'the projection postulate.' It is a solution alright, but requires filling in by some kind of mechanism to be satisfying. We mention a few of these here. The most famous (the one that 'stuck') is the Copenhagen interpretation.

According to the Copenhagen interpretation, the quantum state's function is to provide information about measurement outcomes. Given some preparation of the system in some state, one can then plug this state into the dynamical equations of quantum mechanics to make (probabilistic) predictions about outcomes at later times.

At the root of Bohr's understanding of quantum mechanics was the notion of *complementarity*. This refers to a pair of features or quantities that are both required to understand a system, yet cannot be measured or determined simultaneously. The best known complementary pair are position and momentum of one and the same particle. We saw this earlier: the more we try to pin down one, the less we are able to pin down the other. One could place a screen to detect the interference fringes (and so the wave-like aspects) in a double slit experiment (with no attempt to detect which slit a particle travels through); or one could position detectors at the slits, thus revealing the particle-like aspects. Doing either of these experiments rules out the other: they are mutually exclusive. One can go bonkers trying to figure out ways to evade this embargo and achieve precise determinations of both aspects – but that does not mean it is not a good exercise! The point is that although wave and particle aspects appear to be flatly contradictory, the embargo on simultaneous measurement means that it doesn't cause us trouble: it's wavy when we detect for wavy aspects and particulate when we detect for particulate aspects.

This latter feature points to a kind of relationism in Bohr's interpretation. The exhibition of a property by a system is relative to a measurement designed to reveal it. What Bohr does not want to countenance is the notion that systems 'just have' properties regardless. This relativization is precisely what was behind Einstein's refusal to accept Bohr's approach, remarking that he believed the Moon was there even when nobody was looking (i.e. doing a position measurement). In the case of the cat, it does not have a definite state until a measurement is performed. A 'classical' measurement apparatus is needed to do this: a human eyeball, or a Geiger counter. It doesn't matter so much about the size, but that it can provide an observable output is important.

Hence, in between measurements (when not aligned with some experimental arrangement), the system is described by a quantum state that represents a kind of disposition or potential to generate outcomes. To get the classical world out of quantum mechanics, one needs classical equipment to *register* the results. A meaningful situation, then, demands a system and an apparatus.

One of the most problematic aspects of the Copenhagen interpretation is the special status it assigns to measurement (and the classicality demanded to make sense of it). It almost looks as though the demand for classical measurement devices pulls the rug out from under quantum mechanics, rendering it non-fundamental and non-universal. Just what is so special about those processes we call measurements? This amounts to another form of the measurement problem, though one not grounded in the clash between theory and experience.

Eugene Wigner raised the following problem (known as 'Wigner's friend') for the Copenhagen approach to the Schrödinger cat paradox: what happens if you put the scientist (initially observing the cat in the box) inside an even bigger box, which can be observed by another (exterior) scientist? Now the first scientist is part of the system rather than the measurement device. In this case, the scientist in the box is described by a quantum state, a superposition state (atom + poison + cat + scientist) until a measurement is made. The exterior scientist now has the power to collapse the wavefunction by peeking inside. In principle, of course, this boxing procedure could be carried out *ad infinitum*, producing a sequence of boxes each with their own scientist, like nested Matryoshka dolls, with the most exterior scientist always having the ultimate power to collapse the wavefunctions of those within, which are in an indefinite state until observed.

Wigner found this implication absurd, but proposed instead what strikes many as far more absurd: the *mind* causes quantum states to collapse. So whenever there is a mind, there will be the ability to collapse. Hence, contra Bohr, there is a limit (far smaller than the universe) to the 'system versus apparatus' split enforced by consciousness (awareness). A measurement *result* would seem, after all, to demand awareness of the meaning of the result. Perhaps the cat itself is aware enough to collapse wavefunctions? Such 'mental' intrusions (also espoused by von Neumann) were mercilessly mocked by Bell, who inquired as to what level of mind is needed to collapse a wavefunction: an amoeba or only someone with a PhD?! The problem with this solution, though it is intuitively obvious in a certain sense, is that it introduces another (perhaps deeper) puzzle: what is this mind stuff? How does it work its magic? Wigner has no account, and the solution dissolves into a supernatural speculation. It also leaves the difficult problem of all of that 'pre-mind' time out of which minds themselves must have emerged.

The reasoning in the Wigner's friend example here is, however, perfectly correct, and it basically replays the original Schrödinger cat scenario at a meta-level. If the cat is dragged into the quantum superposition of the poison, which is itself dragged into the superposition of the radioactive substance, then any system that subsequently interacts with the cat (or some of its properties that it can become correlated with) will likewise be dragged into the superposition. This is a simple consequence of the linearity of the dynamics (the evolution described by the Schrödinger equation): superpositions 'infect' systems they come in contact with, like a superposition virus, forcing them into superpositions too. Except they clearly don't, since we know perfectly well we would only ever see a *definitely dead* or *definitely alive* cat. Hence, Wigner's (difficult) decision to let the buck stop at the observing mind.

Though clearly very distinct from the Copenhagen approach, Wigner's proposal too relies on the notion of a primitive measurement process, capable of collapsing the wavefunction (and so destroying any interference that would allow us to see quantum superpositons of distinguishable alternatives). But measurement involves a coupling of physical systems and, unless you're Wigner and believe in a dualist scheme (with mind and matter distinct), must also be governed by the same laws as any other physical interaction. So why does measurement play a special role? Surely it's just physical systems interacting? Why isn't it subject to a quantum description too?

A fairly standard view these days is that measurement is indeed 'just another physical process,' but in real cases of measurement there are some features that make it at least a little special. For one thing, measurement apparatuses are *big*: the detection events in the various experiments we've been discussing (the clicks on a scintillation screen in the two slit experiment; the large spatial separation between the beams of a Stern–Gerlach experiment, etc.) are decidedly macroscopic: they involve large numbers of particles that serve to *amplify* some microscopic event. We know that it is difficult to make large things exhibit quantum interference effects: one can hardly throw a team of scientists through a double slit experiment to build up a wavy interference pattern on a screen (you could never get the ethics approval anyway in these politically correct days)! We find that the large results that our detection equipment indicates are macroscopically distinguishable correlates of the micro-systems they are calibrated to measure.

The backbone of this 'size matters' idea leads us into what has become a standard part of modern quantum mechanics: decoherence. Decoherence has been viewed by some as a sort of magic bullet for the measurement problem, capable of explaining away the absence of interference effects in our observations, while being perfectly within the realm of the physical, and just an application of ordinary quantum mechanics (without projection or any of that funny mind stuff). Recall that the curious interference effects at the root of the double slit experiment and the measurement problem (and entanglement) come about from the existence of the quantum phase θ. If this can be reduced to zero somehow, then the curious phenomena disappear and the behavior looks classical. Decoherence is often touted as one route to the zeroing out of phase, and therefore interference, and therefore the measurement problem.

But it is not a complete miracle cure. The idea is that when interactions occur between a microscopic system (with very few degrees of freedom) and a measuring device (lots of degrees of freedom), we also need to take into account the fact that both are situated in a wider environment with a vast number of degrees of freedom, and evolving according to the (linear) Schrödinger equation with respect to this environment. This environment thus has the effect of 'absorbing' the quantum interference through correlations with these degrees of freedom. The measured system thus engages in a multitude of other interactions with air molecules, photons, and what not, in such a way that in focusing on the apparatus alone one would be forgiven for thinking that it is in a mixed state (i.e. with good old classical probabilities for the alternatives), with no interference (between the alternatives) present at all. In other words, 'for all practical purposes' one has gotten rid of the problem.

Measurement, in this case, can be viewed as converting a genuine superposition into what looks like a mixture (i.e. something that can be dealt with using classical probabilities). In different terms, it looks like an example of an irreversible process, where we have lost information about the superposition into the environment Φ_{env}. But, as the above suggests, the superposition is still lurking in the world, spread throughout all of the degrees of freedom that are involved in any real physical measurement. The interference is still in the total state ($\Phi_{env} \otimes \psi_{sys}$) but not in either of the 'reduced' states: Φ_{env} or ψ_{sys}. To extract the information needed to exhibit the interference would perhaps require a Maxwellian demon, tracking all of the interactions, thus enabling reversibility. But in principle, the process *is* reversible.

This split between quantum and classical is very different from the shifty split of the Copenhagen school, where one moves it depending on context (i.e. on what the system is and what the apparatus is). The effect involved in decoherence is one that occurs by degrees, dependent on the number of degrees of freedom of the environment, rather than on a decision about what the measurement apparatus and the measured system will be. As regards the actual mechanism, it is most frequently expressed as a kind of scattering phenomenon. The environment itself 'measures' the system through scattering. But, as mentioned, given linearity we face a measurement problem (Schrödinger's cat scenario) with decoherence. We might also put pressure on the distinction between 'system' and 'environment' that has a flavor of a Copenhagen-style split about it.

However, having said this, decoherence (or rather its avoidance) it is of great importance in quantum computation. Quantum computation is effective when there is *coherence* between the eigenstates of a system (the $|0\rangle$s and $|1\rangle$s). Hence, it is the interference (in addition to the superpositions, of course) that leads to the exponential (or at least, super-polynomial) speed-up of quantum computation. If this interference is lost then the speed is dampened accordingly. So whether decoherence is able to assist with the conceptual problems of quantum mechanics or not it is nonetheless an important component of the theory.

Hence, we are left with a serious problem: what kind of world do we live in, given decoherence? After all, what it is telling us, if we take it literally, is that superpositions *always persist*, and strictly speaking the appearance of classical outcomes is an illusion. We are, in other words, left with the problem of interpretation of the wavefunction: measurements don't give us definite events, but only the appearance of such. What becomes of the objects and properties that we seem to encounter in everyday life? Are they somehow not the solid, definite entities we thought they were? Many feel that decoherence alone can't work magic, but combined with Hugh Everett's relative-state interpretation, it can resolve the measurement problem, and offer a cogent world picture.

The Everett interpretation was named the 'relative state' interpretation by its original architect, Hugh Everett. It is often presented in terms of 'many worlds,' which involves a curious splitting of the world per measurement made. This was certainly not Everett's intention, which simply involved the self-sufficiency of Schrödinger evolution (in recovering the world, at an observable and unobservable level). Recent work, especially by philosophers, has tended to revert back to a non-splitting conception. However, the reason for the splitting cannot be ignored: it was invoked to introduce a notion of 'happenings' in a quantum world (that is, concrete occurrences rather than superpositions). A purely quantum world governed by nothing other than Schrödinger's equation, with its linearity, would not contain definite events or outcomes. Each branch of a superposition is equally real. According to a re-interpretation of the Everett interpretation, the 'many worlds' interpretation, each branch is a world in itself, so that whenever a superposition occurs (and so branches appear) a real physical branching of the universe occurs. In this way it seeks to square our phenomenal experiences of a single reality of events with the 'bare' formalism of QM with its evident lack of such: since we only experience one world (that branch along which we are thrust, we cannot possibly experience what occurs at the other branches. Staunch defenders of the interpretation, such as David Wallace, don't view Everett's approach as an interpretation at all. Rather, they see it as a more or less *direct* realist reading of the formalism: part of the theory itself.

Aside from the seemingly bizarre nature of the Everett interpretation, there have traditionally been two key problems: (1) the preferred-basis problem; (2) the probability problem. The former problem refers to the fact that we only ever seem to experience well-defined branches (or worlds), not fuzzy worlds corresponding to the interference terms (I_{12} in eq.7.3) involving two well-defined branches simultaneously realized: quantum linearity over the time the universe has existed should surely have generated a blooming, buzzing confusion? So why do we get the nice, classical-looking branches we get? Why do you seem to have such a well-defined location, for example? This is where decoherence theory enters the scene since it

effectively eliminates (or, rather, 'buries') the phases responsible for interference of this kind – it doesn't entirely eliminate the interference and so
doesn't eliminate the preferred-basis problem in anything but a 'practical'
sense; the appearance of definite, classical states is only that: an *appearance*. However, if decoherence is viewed as a scattering phenomenon,
then it faces a preferred-basis problem of its own: in one 'effective branch'
the scattering will go one way and it will go in other ways in other effective
branches. Again, why do we find those nice branches all the time, rather
than strange combinations of the macroscopic alternatives?

The latter refers to the problem of making sense of probabilities (which
surely involves *outcomes*) in a world in which, in a sense, 'everything happens.' It is a little like rather than throwing a dice, we throw the six faces
written down on pieces of paper that all land together: probability 1 every
time! This is to be expected, of course, since Everettians essentially believe
that the Schrödinger equation is pretty much all there is to quantum
mechanics, and that equation is a perfectly deterministic wave equation.
Therefore, the notion of the probability of some event's happening loses
its standard meaning. But we still need to match experimental life. Recent
approaches (the Deutsch-Wallace approach) involve decision-theoretic
derivations of the Born rule linking the wavefunction to experiment. The
idea is that since we don't know where we are in the Everettian multiverse,
with its many branches, or even how many branches there are, we can
derive the probabilities in terms of rational behavior of agents in such a
situation of uncertainty. The probabilities are, in this case, not in the world
but are subjective degrees of belief.

Other issues that are currently being investigated are the extent to which
our world (with its local events and ordinary objects) can be seen to emerge
from the wavefunction – it is generally assumed that there is really just one
universal wavefunction from which everything emerges.

Another approach is the Bohmian interpretation, which is also deterministic, like the Everett approach. Bohmian mechanics supplies particles
with definite positions at all times, but, as mentioned earlier, the interference phenomena come from the addition of a kind of nonlocal field (the
wavefunction) that 'guides' the particles trajectories. Indeed, some many-
worlders claim that since the wavefunction is part of Bohm's approach
(together with the particles in their nice definite configurations) it contains
Everett's approach: all the branching structure is built-in. So why bother
with the particles if you can make do with the wavefunction alone? Of
course, a Bohmian will also face the problem of explaining the quantum
probabilities. In this case it is epistemic, but refers to an ignorance of a
particle's initial configuration.

I think it's fair to say that the Bohmian and Everettian approaches are
the most popular among philosophers in the present day, though there

have been other recent attempts to solve the quantum problems. One notable attempt to ground the projection postulate in a physical mechanism are the 'objective collapse' proposals: the GRW (Ghirardi-Rimini-Weber: the surnames of the interpretation's architects) spontaneous localization approach and the similar gravitationally induced collapse approach of Roger Penrose. They are wavefunction realist, and accept the existence of macroscopic superpositions but show that they are collapsed, by physical mechanisms (a random collapse linked to the number of particles of a system in GRW and gravitational instability caused by superpositions of masses), so quickly that we don't have a chance to observe them. But they don't resolve the measurement problem due to the problem of 'wavefunction tails': they don't completely eliminate the superposition (similarly to decoherence), and so we always face an interpretative problem of explaining what these correspond to in the world.

The measurement problem highlights more than any other conceptual problem in physics how diverse the views of the physical implications of a theory can be. Though most of the experimental and formal details are agreed on, what these *mean* is still all over the place. To return to a point raised in the opening chapter, I say: what is the point of having these theories and performing these experiments if we don't know what they are telling us about the world? To see quite how diverse the interpretative views of physicists (rather than philosophers, though I'm sure we'd find a similar phenomenon), read the recent survey ("A Snapshot of Foundational Attitudes Toward Quantum Mechanics" [43]) carried out by Max Schlosshauer, Johannes Kofler, and Anton Zeilinger, in which a variety of questions about the philosophical implications of quantum mechanics are asked. This should show you that, despite the age of the problem, there is still much to be done to clarify what is going on, let alone in solving the problems.

7.5 Identity Crisis

Recall that the basic feature of statistical mechanics was that states of wholes are given by a kind of averaging over the states of parts. So we must know something about the parts in order to make sense of this. We must know how to count the states of the parts. The crucial question is: given a complex system of particles, how many microstates are there for some macrostate? Here we need to think in terms of *physically distinct* microstates, since we don't want to find ourselves over-counting, and therefore assigning the wrong probability to the macrostate. It turns out that the way we count, and therefore the probabilities for macrostates, depend on whether we are dealing with classical or quantum particles, and with bosons or fermions. Quantum particles have fewer possibilities open to them, and fermions have still fewer possibilities than bosons. Quantum statistics is different to classical statistics: not surprising perhaps, but it is a result that has potentially revisionary implications as regards the metaphysics of objects in the two theories.

There are several features that we quickly need to lay out. Firstly, particles (of the same kind: e.g. electrons) in quantum mechanics have the same 'state-independent' properties: mass, charge, spin, and so on. Such particles are often said to be 'identical' by physicists, but philosophers prefer to call them 'indistinguishable.' They are real identical twins – a puzzling parallelism most probably due to the fact that particles are really excitations of one and the same basic underlying field. They look on the face if it like immediate violations of Leibniz's principle of identity of indiscernibles – something we return to below.

Encountering genuinely identical twins can be confusing, especially when they also dress identically: it is impossible to tell which is which (at least on the basis of appearance alone). This can be used to create mischief of course, since they can be switched without those around realizing – a feature often used as a plot device in films! Elementary particles of the same type are an extreme case of this same phenomenon leading (on a common view) to what is called *permutation invariance* of the laws of quantum mechanics. These twins are so closely matched that not even the laws of physics can tell them apart: no observable quantity can be called upon to tell them apart – in symmetry talk we say that the permutation operation that brings about the switching 'commutes' with all observables, including the Hamiltonian responsible for generating the dynamics. This permutation invariance, itself an implication of the genuine indistinguishability of the permuted systems, is taken to *explain* the difference between classical and quantum statistics. Another explanation is that since quantum particles do not possess definite trajectories (according to some interpretations) they cannot be re-identified at different stages, and so switchings can't be possible.

In terms of plot devices, then, the switching of quantum particles is rather more like the storyline in the film *Freaky Friday* (the Tom Hanks movie *Big* uses the same idea). Here non-qualitative 'personal identities' (memories and souls!) are switched between a pair of bodies, leaving everything qualitative intact – a manoeuvre that clearly involves mind-body dualism. Of course, we might still view 'the soul' as a qualitative feature since it generates actions. If we switched *everything*, including souls (a 'maximal property swap'), then it seems very hard to speak of any real change being effected at all: we've truly left things as they were.

We can put these ideas quite simply in terms of simple worlds; this will enable us to link up to the notion of haecceities discussed earlier in §4.1. Consider a world with just two individuals a and b, and two properties F

and G (not mutually exclusive). There are seven 'worlds' (possibilities) that can be constructed from these few building blocks:

1   $Fa \wedge Gb$
2   $Ga \wedge Fb$
3   $Fa \wedge Gb \wedge Fb$
4   $Fa \wedge Fb \wedge Gb$
5   $Fa \wedge Ga \wedge Gb$
6   $Ga \wedge Fb \wedge Gb$
7   $Fa \wedge Ga \wedge Fb \wedge Gb$

An haecceitist (armed with their 'primitive identities') will hold that 1 & 2, 3 & 4, and 5 & 6 are distinct worlds, while the anti-haecceitist (armed with primitive non-identity or qualitative identity) denies this, seeing just three worlds. The world 7 (a conjunction of 1 & 2) appears to involve a duplicated possibility for the anti-haecceitist, but is kosher for the haecceitist. The haecceitist believes that possibility space is larger than the anti-haecceitist's. The different ways of counting possibilities in classical and quantum statistics corresponds to this way of thinking.

As we have seen, worlds of just this type, differing in what individuals there are and what they are doing, are at the center of many philosophical debates in physics – those based on symmetries. Paul Teller [50] calls these maximal property swaps "counterfactual switching." The scenario requires that the identities of the objects undergoing such swaps is *constant under permutations of their properties*.

In discussing the connections between the Leibniz shift argument and the permutation argument, Teller writes that

> Both problems can be put in terms of a claimed excess of at least apparently possible cases, suggested by the applicability of the tools of reference. These unwanted cases arise by what I will call *counterfactual switching*. In both problems we have names – number-labels of "quantum coins" [binary quantum systems] and number-coordinates of space-time points. In both problems we suppose that there are identity bearing things, the coins or the space-time points, to which these names refer, and that reference is constant across possible cases by supposing shifts of ALL the properties and relations pertaining to one object of reference from that referent to another, so that the new case is utterly indiscernible from the original. The only difference in the cases is taken to be the identity of the underlying bearers of properties and relations. ([50], p. 366)

Teller argues that the peculiarities of quantum statistics force us to abandon the idea that "labels [are] genuinely referring expressions" (ibid., p. 375) – this is an old habit from a classical way of thinking about objects. Instead, Teller adopts a 'quanta' approach. Bosons should be thought of as 'dollars in a checking account' rather than 'coins in a piggy bank': they can be *aggregated* but not counted and distinguished. This can account for the differing probabilities in classical and quantum statistics.

Paul Dirac sums up the basic situation rather nicely:

> If a system in atomic physics contains a number of particles of the same kind, e.g. a number of electrons, the particles are absolutely indistinguishable from one another. No observable change is made when two of them are interchanged. [. . .] A satisfactory theory ought, of course, to count any two observationally indistinguishable states as the same state and to deny that any transition does occur when two similar particles exchange places. ([7], p. 207)

Let's unpack this a little. Firstly, a *permutation* (an 'interchange' in Dirac's terminology) $\hat{P}$ here is simply an automorphism: a reshuffling of the labels of objects without introducing new labels (or an active mapping of the object to another object using some transformation that permutes them – e.g. a rotation). Suppose we have three particles a, b, c. A possible permutation is simply to switch a and c: $(a \to c) \wedge (c \to a)$. Hence, $\hat{P}(a,b,c) = (c,b,a)$. The permutation symmetry of quantum mechanics simply means that the laws are insensitive to such operations: the permutation *group* (containing such switching operations) is a (discrete) symmetry group of quantum mechanics. In the context of quantum mechanics, these maps act as linear operators on a vector space representing the space of possible states of a quantum system.

Now consider the distribution of a system of two indistinguishable particles, 1 and 2, over two distinct one-particle states, ϕ and ψ (states that can be occupied by one particle at a time, where $\phi(1)$ means particle 1 is in the state ϕ). Statistical mechanics is then, very loosely, concerned with the number of ways we can get a distribution of systems (particles) over states: possibility counting. According to Maxwell–Boltzmann counting we get four possibilities (four possible worlds):

$$\phi(1) \cdot \psi(2)$$
$$\phi(2) \cdot \psi(1)$$
$$\phi(1) \cdot \phi(2)$$
$$\psi(1) \cdot \psi(2)$$

This set of possibilities with the assumption of indifference (equiprobability) yields the Maxwell–Bolzmann distribution: each of the four possible states is equally weighted by 1/4. There are two ways of counting in quantum statistics mechanics: Bose–Einstein and Fermi–Dirac. The former gives the following possibilities:

$$2^{-\frac{1}{2}}[\phi(1) \cdot \psi(2) + \phi(2) \cdot \psi(1)]$$
$$\phi(1) \cdot \phi(2)$$
$$\psi(1) \cdot \psi(2)$$

The first possibility here is a superposition state in which the particles are entangled, so that the permuted states possible in the Maxwell–Boltzmann worlds are bundled into a single possibility. We have three possibilities instead of two, each with a probability of 1/3. The latter gives just one possibility (with probability 1):

$$2^{-\frac{1}{2}}[\phi(1) \cdot \psi(2) - \phi(2) \cdot \psi(1)]$$

Again, this is a superposition, but with a difference in sign. There are, then, two ways in which a wavefunction for a pair of particles can change under their permutation: (1) it can remain the same (symmetric); (2) it can change sign (anti-symmetric). This difference, symmetric (or '+') and anti-symmetric (or '–'), is responsible for the dramatically different aggregative behavior of bosons and fermions. Whereas bringing together fermions in the same state at the same location is impossible, there is no problem with bosons. This feature is responsible for the Bose–Einstein condensate phenomena, in which one creates a 'superstate' from many bosons. The identity issues become especially rampant in this context since it seems that, in as much as the whole aggregate is built up from individual bosons, those bosons now share even their relational properties.

We haven't yet said anything about one of the great principles of quantum mechanics implicit in the above: the Pauli-exclusion principle (no two particles can share the same state in a quantum system). This can be seen to be a fairly simple consequence of the anti-symmetry property under the interchange of electrons: when two electrons switch place, the wavefunction changes sign. Recall that in quantum mechanics, what matters is the squared amplitudes, so that changes in sign (positive or negative) will not show up in a physically significant way: the physics will be insensitive to such differences. Now consider a system with two electrons, separated from one another by the distance $x_1 - x_2$. The wavefunction will depend on this separation.

$$\psi(x_1 - x_2) = -\psi(x_2 - x_1) \qquad (7.8)$$

If we then consider the electrons to be at the same point, so that $x_1 = x_2$, then the probability for such a state, as derived from the joint wavefunction for the two electrons, is $\psi(0) = 0 = -\psi(0)$. If an amplitude cancels like this, then it simply means that the state is not possible: it is assigned zero probability. Here we have the apparent 'loss' of classically possible states and the additional loss of 'bosonically' possible states. What is the connection between this difference in possibility counting and the nature and existence of the classical and quantum particle?

The received view is that this signifies a deep difference between classical and quantum particles: roughly, the fact that we count permutations distinct in Maxwell–Bolzmann systems *even though the possibilities are (qualitatively) indistinguishable* implies that the particles have some form of individuality that transcends their properties while quantum particles lack this property (these are the 'haecceities' of course). As Schrödinger puts it:

> [T]he elementary particle is not an individual; it cannot be identified, it lacks "sameness." In technical language it is covered by saying that the particles "obey" a newfangled statistics, either Einstein–Bose or Fermi–Dirac statistics. ([45], p. 197)

I think that some consensus has now been reached that the argument is too quick: the quantum counting can be understood even on the assumption that particles *do* possess non-qualitative haecceities provided one imposes a 'symmetrization postulate' as an 'initial condition' on the quantum state of the composite system formed from the particles. This means that any state will either be symmetric or anti-symmetric, and once in either of these classes it must stay there. Nick Huggett [25] has argued that the argument works in the other direction too: classical mechanics does not imply that classical particles have haecceities, but is just compatible with it. His argument demonstrates that the haecceitistic (unreduced or 'full') phase space (generally associated with classical statistical mechanics and Maxwell–Boltzmann's distribution, with its four element possibility set), leads to a statistical theory that is empirically equivalent to the anti-haecceitistic (reduced) phase space, with its three element possibility set (generally associated to quantum statistical mechanics). He's quite correct; the result simply follows from the equivalence (at the classical level) of the reduced and unreduced phase space descriptions. We simply modify the relative probabilities of the possible states so that a possibility with one particle occupying a distinct state each are twice as likely as those in which two particles occupy the same state. Quantum statistics then simply differs by making the three alternatives equiprobable.

Schrödinger devised some simple examples for making sense of the differences in statistics. Firstly, the difference between Maxwell–Boltzmann counting and so-called Bose-Einstein counting (i.e. the rules for bosons, such as photons), and then on to Fermi–Dirac counting. In each case, rewards are to represent the particles. These come in the form of memorial coins, shillings, and memberships, which are chosen to share some key characteristic with the particles they represent. Then 'Tom,' 'Dick,' and 'Harry' represent possible *states* of the particles: 'two shillings given to Tom' ≡ 'the two particles occupy the same state.'

The specific distribution is determined by the nature of the objects in the examples:

- Memorial coins are distinguishable (fig. 7.6): one is numerically distinct from the other, and a new configuration is generated by exchanging them.
- Shillings are not (fig. 7.7): numerical distinctness, but no new configuration is generated by exchanging them.
- Memberships are singular (fig. 7.8): not possible for two people to share membership or have two identical memberships.

As Schrödinger explains:

> Notice that the counting is natural, logical, and indisputable in every case. [. . .] Memorial coins are individuals distinguished from one another. Shillings, for all intents and purposes, are not, but they are still capable of being owned in the plural. [. . .] There is no point in two boys exchanging their shillings. It does change the situation, however, if one boy gives up his shillings to another. With memberships, neither has a meaning. You can either belong . . . or not. You cannot belong . . . twice over. ([45], p. 206)

Experiment rather than theory points to the different statistics: particles do not seem to behave like memorial coins. This is quite naturally seen to be linked to the nature of the objects themselves: there is 'something funny' about quantum particles that means that they cannot be switched in a meaningful way. This way of thinking – switching or permuting without altering physical states and observables – should put you in mind of the Leibniz shift argument and symmetries. However, as we saw above, multiple options regarding the nature of the objects are possible.

Before closing, let us return to the status of the exclusion principle and the difference between bosons and fermions. Hermann Weyl argued that only the latter particles satisfied Leibniz's principle of identity of indiscernibles, and so counted as individuals in the proper sense:

> The upshot of it all is that the electrons satisfy Leibniz's *principum identitatis indiscernibilium*, or that the electron gas is a monomial aggregate [Fermi–Dirac statistics]. In a profound and precise sense physics corroborates the Mutakallimun: neither to the photon nor to the (positive and negative) electron can one ascribe individuality. As to the Leibniz-Pauli Exclusion Principle, it is found to hold for electrons but not for photons. ([53], p. 247)

The problem Weyl is referring to is, of course, the very different ways that bosons and fermions can form complex systems: the former can be made to share *all* of their intrinsic *and* extrinsic properties, but the latter cannot. This seems to have some genuinely deep impacts on their status as objects if we are using Leibniz's principle. Remember that PII treats an individual

Key:

= Tom   = Dick   = Harry

Fig. 7.6   A representation of the distribution of distinguishable particles (Shakespeare and Newton memorial coins) over a set of states: Tom, Dick, and Harry. There are nine possible permutations of the particles over the states when the particles are distinguishable, as with these coins: situations in which the two particles are shuffled ('*Newton is Tom* and *Shakespeare is Harry*' → '*Shakespeare is Tom* and *Newton is Harry*') are clearly counted as distinct. Note also that both particles can occupy the same state.

as a kind of catalogue of qualitative properties (a bundle view). This won't work here: there are no uniquely distinguishing properties for bosons – even given a strong view of the identity of indiscernibles that includes relational as well as monadic (intrinsic) properties.

We can then understand Weyl's claim from this perspective. Firstly, we cannot distinguish fermions (electrons for example) by their monadic

Key:

Fig. 7.7　A representation of the distribution of *indistinguishable* particles (shilling coins) over a set of states: Tom, Dick, and Harry. There are now six possible permutations of the particles over the states since situations in which the two particles are shuffled ('*shilling is Tom* and *shilling is Harry*' → '*shilling is Tom* and *shilling is Harry*') are no longer counted. However, both particles can still occupy the same state.

qualitative properties, so the ordinary PII wont work here: fermions *are* qualitatively identical in terms of their monadic properties (they all have the same state independent properties, as mentioned). But recall that a pair of fermions will be represented by an anti-symmetric wavefunction: $2^{-\frac{1}{2}}(|\uparrow\rangle_1|\downarrow\rangle_2 - |\downarrow\rangle_1|\uparrow\rangle_2)$. This is spatially symmetric in its parts: any spatial relation that one electron bears, the other will also bear. But this state implies that fermions will always have opposite components of spin in such a system (measurements, say of the z-components of their spins, always find them anti-correlated). We can in such a case say that fermions are 'weakly discernible,' namely by the existence of an irreflexive relation: $R(x, y)$ but not $R(x, x)$ ('has opposite component of spin to'). If we allow the PII to range over these kinds of relational properties then electrons (and other fermions) can be viewed as individuals.

Symmetrized, bosonic states have no such luxury, and the best bet

Fig. 7.8   A representation of the distribution of *indistinguishable* particles (membership – here of the Cambridge Natural Sciences Club) over a set of states: Tom, Dick, and Harry. There are now just three possible arrangements of the particles over the states since situations in which the two particles are shuffled ('*shilling is Tom* and *shilling is Harry*' → '*shilling is Tom* and *shilling is Harry*') are still no longer counted. However, we no longer have the property that multiple particles can occupy the same state (i.e. the exclusion principle is obeyed). [Image source: 'Group portrait of the Cambridge University Natural Science Club, 1890,' The Wellcome Library and The European Library, CC BY-NC (image modified from original.)]

seems to be the kind of view Teller mentions, involving viewing them as quanta. Indeed, Teller suggests that a 'Fock space' description (associated with quantum field theory), which simply tells us 'how many' particles we have in each state resolves the problem. We can also describe fermions in this representation, but each possible state can only be occupied by one or no particles. That they are more 'individual-like' can serve to ground this distinction.

7.6 Further Readings

Our brief discussion has not even scratched the surface of a rich and mature literature on philosophical aspects of quantum mechanics. The suggested readings below offer a reasonably safe and easy path into this literature.

Fun

- Max Schlosshauer, ed. (2011) *Elegance and Enigma: The Quantum Interviews.* Springer.

  - Nice collection of responses to a series of foundational (and not so foundational: e.g. personal) questions about quantum mechanics by some of the central figures in the field.

- David Albert (1994) *Quantum Mechanics and Experience.* Harvard University Press.

  - This treatment of the philosophy of quantum mechanics is about as elementary as it can get while still remaining true to the subject (and not simplifying away too many details). Highly recommended to those struggling with basic mathematical aspects.

Serious

- Rick Hughes (1992) *The Structure and Interpretation of Quantum Mechanics.* Harvard University Press.

  - Solid interweaving of the interpretative issues with the core technical concepts of quantum mechanics. Includes worked examples done in a 'hand-holding' way.

- Chris Isham (1995) *Lectures on Quantum Theory: Mathematical and Structural Foundations.* World Scientific Pub Co Inc.

  - One of the best textbooks in terms of making difficult concepts easy and entertaining. Totally unique in style and approach.

- Leonard Susskind and Art Friedman (2014) *Quantum Mechanics: The Theoretical Minimum.* Basic Books.

  - The perfect book for gaining practical skills in quantum mechanics – a good complement to the two preceding (more conceptually oriented) books.

Connoisseurs

- Michael Redhead (1989) *Incompleteness, Nonlocality, and Realism: A Prolegomenon to the Philosophy of Quantum Mechanics.* Oxford: Clarendon Press.

- In many ways this is the book that heralded the birth of a more rigorous, technical approach to philosophy of physics. Still a must-read for all philosophers of physics.

• Jeff Bub (1999) *Interpreting the Quantum World*. Cambridge University Press.

 - Very systematic discussion of the interpretation of quantum mechanics, focusing on the underlying logical structure of the theory.

• Tim Maudlin (2011) *Quantum Non-Locality and Relativity: Metaphysical Intimations of Modern Physics*. Cambridge University Press.

 - Classic examination of nonlocality in quantum mechanics and its apparent conflict with special relativity.

8 On the Edge:
A Snapshot of Advanced Topics

This final chapter deals with a range of topics at the forefront of research in physics and philosophy of physics: topics you might find yourself dabbling in if you decide to continue in the field. Rather than providing detailed expositions, they are intended to provoke readers into further independent research – indeed, it is impossible to give detailed expositions for some of the examples presented since they are still very new and, in many cases, highly technical. Think of these brief snapshots as mental espresso shots.

The topics chosen are special in a certain sense since they each involve a closing of the gap between philosophy of physics and physics proper (some more than others), often with 'physics-philosopher' collaborations springing up to better understand the nature of the problems. In each case, at the root of the problems is some foundational concept (probability, time, space, causality, computability, etc.) that is up for grabs. They are, then, beyond the mere performing of experiments or churning out of numbers to compare with experiments.

Topics include the possibility of time travel and time machines; the problem of how the structure of the universe can impact on what kind of computations are allowable (including quantum computation); aspects of gauge theories (including the various formulations [with their own ontologies] of electromagnetism and the related Aharonov–Bohm effect); the question of what ontology (particles or fields) is appropriate for quantum field theory; the apparent 'timelessness' of quantum gravity; the application of physics to the 'human sciences'; and the question of why the universe appears to be 'finely tuned' for life (and the related notions of the Anthropic principle and multiverse).

Since these are intended to whet the reader's appetite for future research in the field, suggestions for further readings and also potential research projects are provided at the end of each section.

8.1 Time Travel and Time Machines

Philosophers often distinguish various *grades* of possibility. The two that concern us here are logical possibility and physical possibility. First we need to ask whether time travel is logically possible (or consistent). If it involves no situation in which something both occurs and doesn't occur then we are good to go to the next level: physical possibility. Here we are concerned with whether our laws of physics permit time travel, to see if we might actually be able to construct a time machine.

The standard objections to the consistency of time travel concern the kind of temporal-logical tangles one can get into with a time machine. The grandfather paradox is the most famous of these. But there are also causal loops that appear to allow an effect to be its own cause, 'bootstrapping' itself into existence:

- *Bootstrap Paradox*: this is best seen in the basic timeline in the movie *Predestination* (spoiler follows!) in which a female with both genitalia has a female-to-male sex change, travels back in time as the male and conceives a child with the female version, *then* takes the child back in time to a point at which the child grows into the female version and so into him: hence, we have a case of auto-genesis! Though often baffling, there is no logical contradiction involved in these stories, and so they are logically possible scenarios.
- *Grandfather Paradox*: This involves a situation where you travel back in time to kill your grandfather (on your mother's side, say), but of course if there's no grandfather then there's no mother and so there's no you to go back in time to kill him: auto-infanticide! The film *Back to the Future* involves this kind of timeline, in which Marty travels back to his past and while not killing his mother, almost causes her not to conceive him with his father – with Marty fading from reality as the paradox threatens. But all ends well and Marty returns to his own natural time to happier, wealthier parents. But this kind of story involves a logical contradiction: you (or some fact) both exist and don't exist in this scenario. This kind of thing simply can't happen for the most fundamental of reasons (reality wouldn't make sense if they could), so it looks like time travel and time machines are impossible – no need to assess physical possibility since that depends on logical possibility.

The problem is, the laws of general relativity *do* seem to permit time machines! So what can be going on? At the root of the modern work on time machines (and time travel) are the Einstein equations for gravitation. It was noticed very early on in the history of general relativity that space could be finite but unbounded (closed). Since spacetime is a unified entity in relativistic physics, it prompts the question of whether time can be similarly structured. If space is closed on itself, like on the surface of a sphere, then we can imagine going off in one direction of the space and coming back to the place we set off from without having to turn around. If time can be closed like this, it would mean that we could travel in time and also 'come back' to the place we set off from. But that would correspond to the past! In modern physics having a time machine simply means having a spacetime that has such 'closed timelike curves.' If we had access to such a machine then (assuming we could operate it in such a way that humans can travel along these curves) we could time travel into our pasts in the full-blown science fiction sense.[1]

It is the squishiness of spacetime in general relativity that allows for time machine solutions. While the field of light cones is fixed in special relativity, they can tilt depending on the way energy is distributed in general relativity. Just as one can work from a matter distribution to a spacetime, so one can fix a desirable spacetime (with time-travel-friendly features, such as wormholes) and then figure out what the energy distribution must be like to make it so.[2] A generally relativistic time machine is, then, a spacetime. There is not necessarily a special machine that would take you to the past (though strong gravitational shielding might be needed in some cases). You travel in the spacetime in the usual way, only the spacetime itself has special properties (light cones tilting around in a loop) making it such that your normal travel is taking you into your past. No lightspeed travel is needed and there is no funny dematerialization as with the Doctor Who's TARDIS or H. G. Wells' time machine.

The logician Kurt Gödel discovered a solution (world) of general relativity with the required properties. The solution involves a world with vacuum energy and matter, in which the matter is rotating (and rotating from the standpoint of each location). This causes the light cones to tip so that every single point of the world has a closed timelike curve through it (i.e. starting from any point you could travel into your own past). Gödel linked the physical possibility of universes with closed timelike curves with the idea of a block universe that we discussed in §4.3. His thought is that if one can travel from any point into a region that is past, for any way of slicing the spacetime up into time slices, then the notion of an objective becoming into existence ("objective lapse") makes no sense. Thus he writes, in closing:

> The mere compatibility with the laws of nature of worlds in which there is no distinguished absolute time and in which, therefore, no objective lapse of time can exist, throws some light on the meaning of time also in those worlds in which an absolute can be defined. For, if someone asserts that this absolute time is lapsing, he accepts as a consequence that whether or not an objective lapse of time exists (i.e. whether or not a time in the ordinary sense of the word exists) depends on the particular way in which matter and its motion are arranged in the world1. This is not a straight-forward contradiction; nevertheless, a philosophical view leading to such consequences can hardly be considered as satisfactory. ([20], p. 562)

This seems a curious generalization, from some solutions not having an objective lapse to a statement about time in general, including our own world. However, since the same laws apply to the time machine world and our world the possibility is open, in principle, for implementing those laws to generate the closed curves that would destroy the notion of objective lapse: this possibility is sufficient according to Gödel. To reject his point would involve a direct demonstration showing that our world has features that forbid the generation of such curves.

However, since Gödel's discovery there have been others showing how closed timelike curves can be generated in different ways. All are exotic in some way, though, and unlikely to be realizable in our world.

A serious question remains: how can they be physically possible at all if time travel is logically impossible, as the grandfather-style paradoxes suggest? Stephen Hawking [22] set to work on this problem, and developed what he calls 'chronology protection conjectures' to preserve the logical consistency of the timeline (banning auto-infanticide and the like): the laws of physics always conspire to prevent anything from traveling backward in time, thereby keeping the universe safe for historians.

The philosopher David Lewis [29] suggests a simpler resolution of the grandfather paradox and those like it: since your grandfather did not die, because you are here to prove it, there is no question (given the laws of logic) of your going back in time to kill him. But this might be no better in terms of the possibility of time travel since surely any backwards in time travel will knock variables about leading to contradictions (even if they're unintentional, as in the *Back to the Future* story). That can't happen, so time travel is a precarious business: only contradiction (causal paradox) avoiding trips are possible.[3]

The logical consistency and broad physical possibility of time travel are relatively secure. What remains to be proven is whether *our* universe could permit time travel. We don't think that there are closed timelike curves in it, but whether they might somehow be created artificially in the future is an open problem, though most known methods of creating spacetimes with time machine properties don't seem to be possible for our world.

What to read next

1   Kip Thorne (1995) *Black Holes and Time Warps, Einstein's Outrageous Legacy*. W.W. Norton.
2   Frank Arntzenius and Tim Maudlin (2009) Time Travel and Modern Physics. *Stanford Encyclopedia of Philosophy*: http://plato.stanford. edu/entries/time-travel-phys.
3   John Earman (1995) *Bangs, Crunches, Whimpers, and Shrieks*. Oxford University Press. [A very similar treatment of the relevant Chapter 6 can be found in "Recent Work on Time Travel." In S. Savitt, ed., *Time's Arrows Today* (pp. 268–310). Cambridge University Press, 1995.]

Research projects

1   Have a time travel movie marathon (when you have the time!), and make notes of the timelines employed (using David Lewis' personal and external time idea in [29]), assessing each for both logical and physical consistency: are they bootstrap or grandfather paradox scenarios, or something different?
2   If time travel is possible why haven't we seen time travelers or any evidence of them having visited? Does this provide evidence that time travel is not possible?[4]
3   Do we really need chronology protection theorems?

8.2 Physical Theory and Computability

One might naively think that computation and physics are like chalk and cheese: one (computability) involves abstract stuff (such as computer programs), while the other (physics) is based on physically realized stuff (systems in the world). However, there are a variety of links that can be forged. The Church–Turing Thesis states that a universal Turing machine can compute any function that is comput*able*.[5] Or in other words to be computable is to be computable by a universal Turing machine. The *physical* Church–Turing Thesis is essentially a no-go statement: there is no physically constructible machine (i.e. consistent with physical laws) that can do what an ordinary universal Turing machine (for simplicity, think of a device running an ordinary programming language such as FORTRAN or Python) cannot.[6] Phrased as a no-go claim, the natural instinct of a philosopher-scientist is to look for counterexamples. As such, *Hyper-Computation* is a denial of the physical Church–Turing Thesis: it finds physical scenarios that would allow the computation of functions *beyond* a universal Turing machine's capabilities: there are some physically possible processes that cannot be simulated on a (classical = ordinary) universal Turing machine.

There are two strands to this theme that we shall look at, both concerning the link between physical laws and computability: firstly, the impact of quantum computers (with their apparent speed up relative to classical machines); secondly, the role of spacetime structure (relating back to the supertask-permitting spacetimes briefly mentioned in §5.3).

We briefly mentioned quantum computers in the previous chapter. In 1994, Peter Shor showed how a quantum computer could do in 'polynomial time'[7] what a classical computer could only do in 'exponential time,' namely factoring large integers built by multiplying together pairs of primes (on which is based the RSA encryption keeping your credit card transactions safe). If this is true, then quantum computation (physics, that is) leads to a bigger class of (practically) solvable problems.[8]

This is a slightly distinct claim, certainly related to the physical Church–Turing Thesis, but focussing on *efficiency* instead of the *possibility* of simulation by Turing machines. This states that a Turing machine can compute any function of any physical machine *in the same time* (give or take a polynomial factor). Here, of course, it looks like a quantum computer violates this time-complexity version of the thesis. As Richard Feynman noted [13], a quantum process can't be realized in a system obeying classical rules without requiring exponential time. To simulate these processes, quantum devices are required. In this sense the thesis does indeed appear to be deniable. Physics matters to computation.

For David Deutsch, computation (and the range of solvable problems) also matters to physics and interpretation:

> To those who still cling to a single-universe world-view, I issue this challenge: explain how Shor's algorithm works. I do not merely mean predict that it will work, which is merely a matter of solving a few uncontroversial equations. I mean provide an explanation. When Shor's algorithm has factorized a number, using 10^{500} or so times the computational resources that can be seen to be present, where was the number factorized? There are only about 10^{80} atoms in the entire visible universe, an utterly minuscule number compared with 10^{500}. So if the visible universe were the extent of physical reality, physical reality would not even remotely contain the resources required to factorize such a large number. Who did factorize it, then? How, and where, was the computation performed? ([6], p. 217)

This depends on knowing that there are no 'classical algorithms' capable of matching Shor's algorithm lurking in the periphery, and also knowing that the quantum algorithm is realizable in practice – perhaps reasonable assumptions. Also, there are countless interpretations that 'make sense' of quantum mechanics *without* many worlds, and so also make sense of its computational implications.[9] Christopher Timpson (see 'What to read next,' below) charges Deutsch with committing what he calls a "simulation fallacy": a classical simulation of the factoring algorithm would require many worlds, therefore there are many worlds to perform that computation in the simulated system. As Timpson rightly points out, there is no reason to think that a quantum computer faces the same challenges in terms of required resources.

Deutsch is suggesting that quantum computation's speed-up might play a role in *selecting* a preferred interpretation from a class of what looked like empirically equivalent pictures. While they are empirically equivalent, they are not, according to Deutsch, *explanatorily* equivalent. Since a job of an interpretation is precisely to explain such things as the Shor algorithm there is perhaps more to be said for Deutsch's argument. However, the burden of proof is on Deutsch to demonstrate that the other single-world approaches (given that they are perfectly quantum mechanical) cannot explain the speed-up just as well – this would require, among other things, an account of what is meant by 'explanation,' showing that the account favors many worlds and that we should favor the account of explanation (or that his result is *robust* across all accounts of explanation). That will be a difficult task since the result is a result of quantum mechanics and would therefore be a consequence of any interpretation.

In an interesting development based around 'supertasks' (performing an infinite number of tasks in a finite amount of time), global spacetime structure (according to general relativity) has been linked to the kinds of computation that are possible within a universe. In other words, aspects of spacetime structure impose (and *remove*) limits on what kinds of computation can be carried out.[10] In a similar way, one can simply point to the microstructure of spacetime: if it has the structure of a continuum, then there is, so to speak, a bottomless well of potential information that could be utilized to perform a computation in a way that would defeat a Turing machine. A discrete spacetime automatically reduces what is possible.

The simplest way to see how general relativity might enable computational speed-ups is via gravitational time dilation. This can be seen experimentally in the Hafele and Keating experiment. One can imagine a kind of twins paradox using this setup in which one twin (Angelina) performs a computation in a region of stronger gravity than the other (Brad), who will wait patiently for some computation to be performed. What is needed, specifically, is a spacetime in which Angelina's world-line with infinite proper length (i.e. with infinitely many tick-tocks on her wristwatch) lies in or on Brad's past light cone. Angelina can then perform the computation (in infinite time) and transmit the result to Brad (who receives it in finite time) – one can try to get similar 'hypercomputation' results with radical accelerations of the computer, which would undergo a round trip journey.

Such spacetimes sound odd, but they do exist as possible solutions (worlds) of general relativity – they are called "Malament–Hogarth spacetimes." Just as with the time travel scenario above, the computer that Angelina uses doesn't need to be special in any way; it is the spacetime structure that is special, allowing supertasks (or hypercomputations) to be performed: the spacetime 'slows down,' rather than the device speeding up.

As with the case of quantum computers, there is a question of actually physically realizing these theoretical devices as computers (with memory registers and the like). One can't access the superpositions of quantum computers directly, and it seems even more difficult to see how the spacetime continuum could be channeled into computer construction (of physical computers). However, it is hard to come up with definitive objections to the physical possibility of such hypercomputations, and so it remains a controversial subject.[11]

The above ideas exploited the structure of the universe to tell us something about computation. But we can also consider the other direction: exploiting computation to tell us something deep about the structure and nature of the universe. This has been a more recent strategy, with claims that the world 'is made of information' becoming fairly common. The buzz phrase is John Wheeler's 'It from Bit': the furniture of the universe is fundamentally informational. However, the obvious problem with trying to make ontological sense of this is that information is *abstract*, something that depends on realization *in* a physical system. Hence, how can it possibly be considered physically fundamental? However, in Wheeler's scheme things were not quite so simple, and there was an interplay between ontology and epistemology (the latter playing a role in the former) grounded in a notion of a 'participatory universe.' This is itself grounded in a special status accorded to observation (in terms of creating phenomena) and the choices involved in what to observe. For Wheeler, in order to bring about physical reality, one needs an observer to make a measurement – this is understood in terms of asking Nature a 'yes/no' question (hence 'bit'). In other words, 'bit' for Wheeler was not abstract, but was certainly subjective – see [55]. So the standard objection to theses about reality being in some sense information doesn't quite apply to such a scheme. However, at the same time, this alternative approach doesn't quite mesh with what we ordinarily mean by information. In any case, such proposals, given that they link epistemology and fundamental ontology, offer ripe pickings for philosophers of physics.

What to read next

1 Scott Aaronson (2013) Why Philosophers Should Care about Computational Complexity. In B. Jack Copeland, Carl J. Posy, and Oron Shagrir, eds., *Computability: Turing, Gödel, Church, and Beyond* (pp. 261–327). MIT Press.
2 John Earman and John Norton (1993) Forever is A Day: Supertasks in Pitowsky and Malament–Hogarth Spacetimes. *Philosophy of Science* 60: 22–42.
3 Chris Timpson (2008) Philosophical Aspects of Quantum Information Theory. In D. Rickles, ed., *The Ashgate Companion to Philosophy of Physics* (pp. 197–261). Ashgate.

Research projects

1 Can one satisfactorily respond to Deutsch's challenge to explain Shor's factoring algorithm *without* invoking many worlds?
2 How much information is in a qubit?

3   How serious is the threat to the physical Church–Turing thesis posed by Malament–Hogarth spacetimes?

8.3 Gauge Pressure

We saw in §4.4 that coordinates in general relativity do not have quite the same meaning as in pre-GR theories. They can serve to label points, but not in a way that latches onto 'real spacetime points.' Gauge is a more general way of speaking of coordinates in this sense. One can label other things than bits of space, and in the same way these labels are often devices set up for a more convenient description.

Let's give an easy example to get us going. Suppose I'm running a 'biggest vegetable' competition and need to determine the winner from a pair of marrows. Clearly what we need are the differences between the starting points and endpoints of the marrows: we measure *differences*, not any absolute values. This means that even if I only had a broken tape measure, which started at 15 cm, I could still take a perfectly good measurement with this. What matters is *end – origin* (where *end* > *origin*). Hence, the numerical values we might assign don't have any physical significance: we could each use a distinct set of numbers (e.g. inches rather than centimetres), and agree on the winning marrow. The numbers on a tape measure are a standard example of a gauge. There is quite clearly also a conventional element involved: we could use miles if we so desired, but it would be cumbersome to deploy for such a small object as a marrow (even for a prize-winning marrow).

The notion that we can speak freely of the *same* physical quantity (having the magnitude of length) using many different labeling conventions is a kind of gauge-invariance. This is clearly a rather trivial example, but more interesting examples can be found in which the laws (and quantities) of a physical theory are invariant under more interesting transformations than conversions between units of length measurement. The hole argument featured one example, in which diffeomorphisms (a very general topological transformation) take the place of the changes of units and the laws and observables of the theory take the place of marrow measurements. The group of such transformations is known as the 'gauge group' of the theory. The invariance of some items relative to elements of this group is known as 'gauge symmetry' – and the ability to arbitrarily use any member from an equivalence class of states related by a gauge symmetry is known as 'gauge freedom.'

For example, one could develop an electromagnetic potential's value in many ways off an initial hypersurface, but without any empirical differences in the evolved states. To see this, note that the electric and magnetic fields are related to the 'scalar' and 'vector' potentials as follows:

$$\mathbf{E} = -\text{grad } \Phi - \frac{\partial A}{\partial t}$$

$$\mathbf{B} = \text{curl } A$$

Writing the fields in this form makes certain calculations easier. But in this form we face an underdetermination of the vector potential A by the magnetic field. Since $A = A + \text{grad} f$ (for smooth functions of spacetime coordinates, f), it follows that $\text{curl} A = \text{curl}(A + \text{grad} f) = \mathbf{B}$, and so many (formally) distinct vector potentials will represent the same magnetic field (since the curl of a gradient is zero).[12] The transformation from one vector potential to another, $A \rightarrow A + \text{grad} f$, is an example of a gauge transformation, where A is known as the *gauge field*. Physical quantities will be independent of this field (and dependent only on the magnetic field, or the vector potential 'up to an arbitrary gradient'; that is, 'up to a gauge transformation') and the laws will be covariant with respect to transformations between gauge-related vector potentials. Physical quantities and laws are simply blind to such transformations.

Eugene Wigner describes the introduction of these (seemingly impotent) potentials to represent the states of the electromagnetic field as follows:

> In order to describe the interaction of charges with the electromagnetic field, one first introduces new quantities to describe the electromagnetic field, the so-called electromagnetic potentials. From these, the components of the electromagnetic field can be easily calculated, but not conversely. Furthermore, the potentials are not uniquely determined by the field; several potentials (those differing by gradient) give the same field. It follows that the potentials cannot be measurable, and, in fact, only such quantities can be measurable which are invariant under the transformations which are arbitrary in the potential. This invariance is, of course, an artificial one, similar to that which we could obtain by introducing into our equations the location of a ghost. The equations then must be invariant with respect to changes of the coordinate of the ghost. One does not see, in fact, what good the introduction of the coordinate of the ghost does. ([56], p. 22)

When we have a theory with a gauge group, like electromagnetism, we find that our initial (Cauchy) data made at some instant, no matter how complete, will not serve to fix the physical situation if we understand the gauge variables (or frame) to be responsible for representing our physical frame. This distinguishes gauge freedom from the kind of situation we found in the Leibniz shift examples. That also involved an underdetermination of the physical state by the laws, but was not sufficient for indeterminism of the kind found here. Indeterminism, in the gauge theory sense, requires that *no amount* of specification of initial data can secure unique future values for some physical quantity or object. The gauge transformations can be performed *locally*, so that we can choose to do it *after* some initial specification of values has been made. In the case of Newtonian mechanics, there are enough evolution equations to allow us to evolve (or 'propagate') all of the physical magnitudes once a labeling of the particles and localizations and velocities have been settled on since the symmetries there are *globally* defined: once we 'break' them, so to speak, everything is fixed thereafter (even if the choice was arbitrary to begin with).[13] The indeterminism that lies at the core of gauge theories is an underdetermination of solutions of the equations of motion given an initial data set: no amount of specifying in this sense will enable the laws to develop the data into a unique solution. It is, then, rather a more serious epistemological defect than the absolute velocities in Newton's theory.

You might wonder why physicists play around with gauge freedom if it's both damaging and inconsequential. That's worth wondering about, there's still much work to be done in fully figuring this out. But we can say that gauge symmetries play a crucial role in identifying the physical content of theories; namely, as that which remains invariant under the gauge transformations. This sounds similar to symmetries, of course, and they are related, but note that gauge symmetries are often referred to as 'redundancies,' which indicates a crucial difference. Simply put, symmetries in general are structure-preserving transformations. Usually, symmetries relate one solution of the equations of motion of some theory (or a physical system described by the theory) to another *physically distinct* solution, albeit in a way that preserves some features (such as the laws or observables). In other words, symmetries are transformations that keep the system's state within the set of *physically possible states*. The orbit under the action of a symmetry group consists of points representing *distinct* situations related by the symmetry transformation. With gauge symmetries, however, though the transformations do still map between physically possible states, they are viewed as representations of *one and the same physical state* (this protects physical determinism, of course, since there are no physically distinct alternate possibilities). The orbit of the gauge group consists of points that are not just identical in certain physical respects (i.e. in terms of what they represent): they are physically identical, period. In fact, 'physical state,' in this context, is really just shorthand for 'equivalence class of states under gauge symmetries,' so that physical states are represented by entire gauge orbits rather than their elements.

While symmetries can result in physically distinguishable scenarios, gauge redundancies (as the name suggests) result in no physically observable differences, and so, to get at the 'real structure,' are usually removed by a 'quotienting' procedure that leaves one with a (reduced) space of orbits of the gauge group. Hence, various representations of a theory's content can be given, but they will be considered to lie in an equivalence class such that each represents the same physical state of the world. For this reason, Richard Healey adopts the stance that gauge symmetries are cases of 'multiple realizability': many ways of realizing some physical state (albeit with redundancy involved in the representation). But the realizations are generated by surplus formal structure (part of the mathematical representation of a theory that should not be given any ontological weight as it stands): "Understood realistically, the [gauge] theory is epistemologically defective, because it postulates a theoretical structure that is not measurable even if the theory is true" ([23], p. 158). As Healey points out, and as we saw above, the natural course of action is to treat the realizations as representations of the same physical situation. Indeed, this 'mathematical surplus' approach is the default.

I'm belaboring the point here since one might be led to think that the obvious ontological option in the case of electromagnetism is to commit to the reality of the electric and magnetic fields only: these are what we measure (potentials are *unmeasurable*), and these are what remain invariant under the gauge transformations (potentials are shifted: gauge-variant). This gauge-invariant interpretation is indeed an obvious choice: we get a one-to-one mapping between the fields and the world (modulo the usual idealizations involved in any physical theory). We simply take the vector potential to be *unphysical*. However, when we include quantum mechanics the situation appears to radically change, with the vector potential playing a role in the dynamics – the potential figures in the Schrödinger equation and so despite its 'gaugey' nature, might need to be given a physical reading. But the problem is that the gauge field is as (directly) unobservable as ever it was: it is determined by the equations of motion only up to the addition of the arbitrary gradient of a function of spacetime.

An experiment devised by Yakir Aharonov and David Bohm (describing the Aharonov–Bohm effect, often just called the 'AB-effect') appears to breathe life into the vector potential. The experiment is much like the double slit experiment, only a solenoid (a coil in which a magnetic field can be turned on and off) stands behind the slits. A beam of electrons is fired at a detection screen, and one finds the familiar interference pattern. The novelty is that one can cause the interference pattern to shift (i.e. there is a phase shift), in a predictable way, by turning the current in the solenoid (and so the magnetic field) on and off. That might not strike you as particularly groundbreaking – switching a field on and off is clearly causing it: what's the problem? The curious aspect is that the magnetic field is *zero* in the path of the electrons (outside of the solenoid), so this should not be happening. The proposed reality of the vector potential then turns on the fact that this is *non-zero* in the path of the electrons – due to the fact that integrating the vector potential around a closed loop is equal to the magnetic flux the loop encloses.[14]

We could stick with the magnetic field as fundamental, but to do so would now involve viewing it as acting-at-a-distance on the electrons. We invoke the potential because we have a desire for local action in physics. However, with the vector potential interpretation we face the problem of underdetermination in which we cannot say *which* of an infinity of potentials is responsible for the phase shift: the dynamics can only determine the evolution of potentials up to a gauge transformation. Hence, we have to trade local action for indeterminism. How to decide?

Note that the vector potential is still not measured in the experiment (how could it be?), but the loop integral mentioned above ($\oint A \cdot dx$, known as an 'holonomy': carrying the vector potential around a loop) is, and this will give the same value for any of the gauge-related potentials: it is gauge-invariant, like the magnetic field. This suggests using such holonomies as an alternative interpretation, which both avoids the problem of action-at-a-distance and the indeterminism/non-measurability: the best of both worlds! The holonomy embodies the invariant structure of the vector potentials since many vector potentials are subsumed in one and the same holonomy: vector potentials give the same holonomies when they are gauge related. However, it is a little hard to see these loopy entities as the stuff of physical reality. Although determinism and local action are restored, another kind of nonlocality re-enters, though of a rather curious kind. The nonlocality concerns the 'spread outness' of the variables: they are not localized at points but rather are represented in a 'space of loops.'[15] But must we be committed to this space in an ontological space? If we are so forced, then the problems of action-at-a-distance and indeterminism look rather less bizarre by comparison. But there are always choices that can be made about which parts of a mathematical structure should be mapped to the world.

To sum up: it turns out that there are several distinct ways to formulate gauge theories, each pointing to a distinct interpretation (a different kind of world, with different fundamental furniture), and having distinct virtues and vices. In the electromagnetic case (though it generalizes easily to other gauge theories), you can use **local** quantities, such as potentials A_i, to represent what's going on; or you can use **nonlocal** entities such as magnetic fields; or you can use **holistic** entities such as the integral of A_i around a path enclosing the region with the magnetic field. The problem with using potentials is that, though they allow a nice local-action account of the Aharonov–Bohm effect, they are 'unphysical' variables: not measurable or predictable by the laws. The problem with using magnetic fields is that though they are measurable and predictable, they must act at a distance in order to explain the Aharonov–Bohm effect (well tested and shown to occur). The problem with using holonomies is that, though they are also gauge invariant, they appear ontologically suspect since they are 'spread out' (a different kind of nonlocality). Given that this set of options is pretty much generic for gauge theories, it is a pressing task for philosophers of physics to figure out which is to be recommended, or say why the choice doesn't matter. Crucial to the pros and cons of these interpretive options is the Aharonov–Bohm effect, but it doesn't necessitate that we believe in the reality of gauge potentials (aka 'the traditional physicist's view').[16]

What to read next

1 Michael Redhead (2003) The Interpretation of Gauge Symmetry. In K. Brading and E. Castellani, eds., *Symmetries in Physics: Philosophical Reflections* (pp. 124–39). Cambridge University Press.
2 Gordon Belot (1998) Understanding Electromagnetism. *British Journal for the Philosophy of Science* **49**(4): 531–555.
3 Dean Rickles (2008) *Symmetry, Structure, and Spacetime.* Elsevier.

Research projects

1 Work out what in your opinion is the best ontology for electromagnetism: fields, potentials, holonomies, or something else. Answer with reference to 'locality (separability) versus holism (non-separability)'; 'locality versus action-at-a-distance'; and 'determinism versus indeterminism.'
2 Should the behavior of the electromagnetic field with *quantum* particles have any significance for our interpretation of the purely classical theory?
3 Why do we set theories up with gauge freedom? Why not do everything *without* adding redundant, unphysical elements?

8.4 Quantum Fields

Quantum field theory is the natural home for describing the interaction of charged point particles via the electromagnetic, strong, and weak fields: three of the four forces of nature (gravity not included). In this approach the discrete, particlelike 'excitations' (photons, gluons, etc.) of the various (continuous) fields *mediate* the various interactions – photons, for example, mediate the electromagnetic interaction between charged particles in the context of quantum electrodynamics, providing a nice local account of how such interactions happen. Given that three interactions have succumbed to the quantum field approach, it makes sense to try to accommodate gravity in the same formal framework: here the gravitational interaction would be mediated by the 'graviton' (technically: a massless, spin-2 quantum particle of the gravitational field). Before we turn to gravity, let us first say something about quantum field theory itself.

As with gauge theories, much of the philosophical work on quantum field theory has tended to focus on what the basic ontology is. We face a difficult problem here, as with gauge theories, since there are many formulations of quantum field theories too, which recommend distinct ontologies. For example, if we use what is called the 'occupation number' representation (using Fock space: Hilbert spaces tensor-producted together), then we can make some sense of there being particles (or rather 'quanta') in the theory, though, as our earlier discussion of particle statistics implies, they are not quite what we usually mean by particle: though we can say *how many* we have, they cannot be individually labeled. But, from what we have said above, if there can be physical states with no particles present (the vacuum), then it is hard to view particles as fundamental. There are also problems in upholding a particle picture in other quantum field theoretic contexts, such as curved spaces in which the notion of particle is seemingly impossible to establish (in the absence of the symmetries supplied by the usual spacetime metric) – to speak of particle number as an observable quantity requires that there be a particle number *operator*, but this seems to be specific to flat, Minkowski spacetime.

A field approach (i.e. in which fields are the basic entities in the world) is the natural alternative to a particle interpretation. But there are issues here too. Not least the problem of mapping quantum fields to the world. Particles can at least be directly associated with objects in the world, but fields in a quantum theory are operator-valued, so we have an inherent indirectness in the mapping. Attempts have been made to reconnect fields and world by employing *expectation values* of the field instead, which provides a numerical value. But such approaches still involve an indirectness since 'expectation' is a statistical concept invoking the average value given many measurements. Moreover, in most practical applications of quantum field theory it is the particles that are center stage – in scattering experiments at CERN, for example.[17]

Note that these ontological considerations go deeper than the usual interpretative options for quantum mechanics (many worlds, Copenhagen, etc.): whatever basic ontology we settle on will *then* face this next level of interpretation. Likewise, if we find that the world is described by quantum strings or branes, then these too will face the usual interpretive suspects for quantum theories.

Ordinary quantum mechanics and quantum field theory are brought closer together if the wavefunction is considered the primary thing. In the case of quantum field theory this is more properly a wave*functional*, or a function that takes another function (in this case the field associating field values to spacetime points) to some number (the probability for finding some classical field configuration). The usual measurement problem arises given this wavefunctional approach since we can have superpositions of classical field configurations (just as we could have superpositions of, e.g. spin states of a single particle). This also faces problems. Again, it's hard to square with the apparent 'practical primacy' of particles. Moreover, the interpretive problems associated with the measurement problem take center stage. And, perhaps worse, the wavefunctional is hard to make sense of given that it maps entities defined on spacetime, but itself is defined with respect to a very complicated field-configuration space.[18]

There are very many more philosophically interesting issues facing quantum field theory. To delve into these would require too much by way of preparation.[19] Instead, let us simply focus on one aspect that stems from the relative treatments of vacuum in quantum field theory and general relativity, and then segue into a discussion of quantum gravity, which involves an apparent failure of the usual techniques of quantum field theory when applied to gravity. Many of the problems have to do with the treatment of space and time in the two contexts.

For example, in the good old days 'vacuum' meant complete emptiness: no energy, no matter, no particles, no fields. Just pristine empty space. Both quantum field theory and general relativity spoil this clean division into empty space and matter. In our discussion of the hole argument, we saw how space becomes more 'matterlike' in general relativity, obeying its own equations of motion, allowing 'ripples of spacetime' (gravitational wave solutions) to be used to do *work*. Mixing quantum mechanics and special relativity (to give quantum field theory) spoils things in a different way since the *vacuum state* is no longer a state with zero energy, but merely a state with the 'lowest' (non-zero) energy. As such it is distinguished in being the ground state, but not really so different from non-vacuum states ontologically speaking.

This framework leads to a serious conflict between general relativity and quantum theory known as the 'cosmological constant problem.' The energy spectrum for an harmonic oscillator, $E_N = (N + 1/2)\omega$, has a non-vanishing ground state in quantum mechanics (the 'zero-point energy') resulting from the uncertainty principle (so that it is impossible to fully 'freeze' a particle so that it has no motion at all, no oscillations). In quantum field theory the field is viewed as an infinite family of these harmonic oscillators (one at each point of space, and each with a little bit of zero point energy), so that the total energy density of the quantum vacuum must be infinite. Of course, general relativity couples spacetime geometry to energy, but if this energy is infinite in any spacetime region then that region will have an infinite curvature. But it clearly doesn't: so what is wrong?

There are ways to tame this infinity. For example, just as there is no absolute voltage in classical electromagnetism (because there is no zero point, only potential differences between points), so one can say the same about the quantum vacuum. If only energy *differences* between the vacuum and energy states make any sense we can simply rescale the energy of the vacuum down to zero. The absolute value of the energy density of the quantum vacuum is unobservable; the value one gives is largely a matter of convention.[20]

But, the clash between general relativity and quantum field theory strikes back since this rescaling is not possible in cosmology. The cosmological constant, or vacuum energy, is taken to just be a measure of the energy density of empty space. This allows an experimental approach to be adopted to determine the correct value using general relativity's link between mass-energy and spacetime curvature. The energy density ρ, as actually observed in curvature measurements, comes out at around 10^{-30}g cm^{-3}: very close to zero, which is a long way from the naive (enormous) expectations suggested by quantum field theory – that is, once we perform various other infinity-taming procedures, that we needn't go into here. Resolving this problem is one of the challenges for a quantum theory of gravity. A straightforward application of the usual rules of quantum field theory does not work with gravity – though it has taken many decades of struggle to figure this out.[21]

This is just one example of the conflict between our two best theories of the universe, general relativity and quantum field theory. There are others, most often stemming from the different ways space and time are treated in these frameworks: in quantum field theories, space and time function as fixed structures, seemingly essential for making sense of the core components of the theory (the inner product, the symmetries, the probability interpretation, etc.); in general relativity, spacetime is a dynamical entity.

As in the other examples of quantum field theories (especially for the strong interaction, which has certain important features in common with gravity), quantum gravity along these lines runs into formal difficulties: when one tries to compute probability amplitudes for processes involving graviton exchange (and, indeed, involving the exchange of other particles in quantum field theory) we get infinities out. Since we cannot readily make sense of measuring infinite quantities, something has clearly gone wrong. In standard quantum field theories, quantum electrodynamics for example, these infinities can be 'absorbed' in a procedure called *renormalization* (similar to the rescaling of the zero point energy as mentioned above). The problems occur when the region that is being integrated over involves extremely short distances (or, equivalently, very high momenta, and so energies, for the 'virtual particles').[22] In quantum gravity, as the momenta increase (or the distances decrease) the strength of the interactions grow without limit, so that the divergences get progressively worse – the problem is: gravity *gravitates*, so that the gravitons will interact with other gravitons (including themselves!). The usual trick, in quantum field theory, of calculating physical quantities (that one might measure in experiments) by expanding them as power series (in the coupling constant giving the interaction's strength) fails in this case, since the early terms in such a series will not provide a good approximation to the whole series.

This gives us one way of understanding quantum gravity: a *micro-theory* of gravity. Or, since gravity is understood via the geometry of spacetime, the microstructure of spacetime geometry. (Hence, quantum gravity is often understood as a theory of quantum spacetime. But there are approaches that do not involve quantum properties of the gravitational field (and so spacetime); they do not *quantize* gravity at all, but might, for example, have gravity (and spacetime) emerge in a classical limit.) At large scales we know general relativity works well, but it buckles at short distances (with singularities being one symptom of this). The scales at which infinities appear like this signal the limit of applicability of general relativity, and point to the need for a quantum theory of gravity (or some other theory that can reproduce the predictions of general relativity at those scales in which it *does* work – superstring theory is an example of this approach). In other words, the picture of a smooth spacetime that we are used to in physics might fail at a certain energy. (This limit of applicability of the general theory of relativity is known as the *Planck scale*. Quantum gravity is, for this reason, also sometimes understood as whatever physics eventually describes the world at the Planck scale.) This in itself points to a more skeptical attitude towards the techniques of (orthodox) quantum field theory since they assume a spacetime that is smooth and continuous 'all the way down.' There are recent approaches that distance themselves from this assumption, either putting discreteness in as a basic postulate (as in 'causal set theory') or else deriving discreteness as a consequence of the theory (as in 'loop quantum gravity').[23] Without the assumption of smoothness, many of the divergence (i.e. infinity) problems evaporate since there is an end point to the scales one can probe (that is, there is a smallest length, area, and volume).[24]

Much of the discussion, philosophical and technical, tends to deal with *free* quantum fields, in which interactions are 'turned off.' Dealing with interactions is notoriously difficult and, indeed, faces something known as 'Haag's Theorem,' which indicates that interacting quantum field theory cannot be established within a consistent mathematical framework. This is suggestive of the difficulty of quantum field theory, which is justified in many ways. To discuss its conceptual issues properly requires deep and lengthy investigation: there's no way around it. I can do no better than to point readers at the beginning of this journey to Paul Teller's book in the 'What to read next' section below.

What to read next

1   Paul Teller (1995) *An Interpretive Introduction to Quantum Field Theory.* Princeton: Princeton University Press.
2   Sunny Auyang (1995) *How is Quantum Field Theory Possible?* Oxford University Press.
3   Meinard Kuhlmann, Holger Lyre, and Andrew Wayne, eds. (2002) *Ontological Aspects of Quantum Field Theory.* World Scientific Pub Co Inc.

Research projects

1   Is quantum field theory, fundamentally, a theory of particles or fields (or neither)?
2   What exactly is it about curved spacetime that causes problems for a particle interpretation of quantum theory? Can particles be included at all in such curved backgrounds? Where does this leave particles in a quantum theory of general relativity?
3   Describe and evaluate wavefunction realism in the context of quantum field theory.

8.5 Frozen Time in Quantum Gravity

There are very many points of contact between physics and philosophy in the case of quantum gravity. Here, to keep the length manageable, we focus on an issue that has received the most attention from philosophers, achieving a degree of notoriety: the problem of time – it helps that this has much in common with the hole argument discussed in §4.4.

The response to the hole argument according to which diffeomorphism invariance points to gauge freedom in the theory (understood relationally or otherwise), so that diffeomorphisms do not correspond to physical changes, has some curious consequences: since time evolution is an example of such a diffeomorphism transformation (e.g. mapping some point (1,0,0,0) into (0,0,0,0)), the observables of the theory – the gauge-invariant (diffeomorphism-invariant) content – are insensitive to such changes. Indeed, the 'temporal' evolution takes place along a gauge orbit so understood since each temporal diffeomorphism counts as a gauge transformation. But surely a theory's observables are the kinds of thing that change over time, from one instant to the next? If not, then general relativity appears to be *frozen*, with observables that look the same on each time slice through the spacetime. This doesn't correspond to our experimental and experiential life, so mustn't we reject the theory?

Note that there is an interesting link to relationalist and substantivalist conceptions of spacetime here. According to the logic of Earman and Norton's presentation of the hole argument, the latter will view each of the temporal advances as generating physically distinct scenarios (since the observables are located on different time slices in each case), but there will nonetheless be no observable change: all invariants are preserved, just as the formally distinct futures in the hole argument preserved invariants (and observable content). Thus, the substantivalist faces the problem of the hole argument in this temporal case: possibilities that differ only in which points (now characterizing a time slice) play which role. But the relationalist's standard Leibnizian trick of collapsing all the substantivalist's indiscernible possibilities into one physical possibility doesn't help us here, since it leaves us with data on a *single* three-dimensional slice. Any attempt to push the data forward will generate an unphysical possibility, so we appear to be stuck on this slice, unable to generate change! It doesn't matter whether we are substantivalists or relationalists here: there will be no change in observables if those observables are defined relative to spacetime (manifold) points.

The problem takes on its clearest form in the *canonical* formulation of general relativity, in which the dynamics (time evolution) of the geometry of space is generated by a Hamiltonian function: we find that it vanishes (as is expected given its gaugey nature: it can't push the data forward in a non-unphysical way). Being a gauge theory, the Hamiltonian is really to be understood as a family of constraints: three spatial (or 'diffeomorphism' or 'vector') constraints and a temporal one (known officially as the 'Hamiltonian' or 'scalar' constraint). We can think of the spatial constraints as simply transformations that bring about ordinary coordinate changes on a spatial slice (the kind found in the hole argument). These can be dealt with in the usual 'equivalence class' manner: the transformations are unphysical → the physical stuff is the equivalence class → all is well. The Hamiltonian constraint generates displacements of data *off* the spatial slice. The overall Hamiltonian, responsible for generating the time evolution of states in the theory, is simply a linear combination of these two kinds of constraint. Taken together, they generate (infinitesimal) spacetime diffeomorphisms – this explains why physical observables (sometimes called 'Dirac observables') can't evolve with respect to the Hamiltonian, since such observables are gauge-invariant entities and diffeomorphisms are gauge transformations.

This problem infects quantum gravity in a fairly direct way: since the Hamiltonian and the observables are quantized by turning them into operators (and imposing commutation conditions), the problem sticks around.[25] But it takes an even worse form: whereas the Schrödinger equation usually involves a t against which quantum states evolve (i.e. as $\hat{H}\Psi = i\frac{d}{dt}\Psi$), we have a truly timeless theory now: the t doesn't even make an appearance in the theory since the classical Hamiltonian, to be quantized and converted into the Schrödinger equation (or rather its gravitational analogue, 'the Wheeler–DeWitt equation'), vanishes: $\hat{H}\Psi = 0$. It is a sum of constraints (on initial data) rather than a true dynamical equation. But without a time, how do things change and evolve? How is there motion?

We can choose to deal with this problem at either the classical (pre-quantization) level or the quantum level. In either case, we have two broad options: either *find* a time buried in the formalism or else bite the bullet (that reality is timeless and changeless), leaving the task of explaining away the appearance of time and change in the world (as a kind of illusion). The physicist Karel Kuchař suggested the terminology of 'Parmenidean' and 'Heraclitean' to describe these fundamentally 'time-full' and 'timeless' responses.

Kuchař himself is Heraclitean, and sees as the only way out of the frozen formalism (and the quantum version) a conceptual distinction between the constraints so that only the diffeomorphism (spatial) constraints are to be viewed as generators of gauge transformations. This leaves the problem of finding a hidden 'internal' time (and so a physical, genuine Hamiltonian) buried among the phase space variables against which real observable change happens. This is a difficult issue that we can't go into here, but we can point out that some responses along these use the distribution of matter to 'fix' some coordinate system, eliminating the gauge symmetry in the process. The usage of physical degrees of freedom to resolve the problem is, in general, the most promising way to go.

Perhaps the most philosophically shocking approach is Julian Barbour's Parmenidean proposal. Barbour argues that we should accept the frozen formalism: time really doesn't exist! What exists is a space of Nows (spatial, three-dimensional geometries) that he calls 'Platonia.' In the context of quantum gravity we would then have a probability distribution over this configuration space (as determined by the Hamiltonian constraint or Wheeler–DeWitt equation) that delivers an amplitude for each possible Now – where each three-geometry is taken to correspond to a 'possible instant of experienced time.'

But unlike the Heraclitean proposals, which strive to get out *real* change (and so can account for our experience of a changing world), on Parmenidean proposals there remains the problem of accounting for the appearance of change. Barbour attempts to resolve this puzzle by appealing to what he calls 'time capsules,' patterns that encode an appearance of the motion, change, or history of a system – he conjectures that the probability distribution determined by the Wheeler–DeWitt equation is *peaked* on time capsules, making their realization more probable (this makes it more likely we will experience a world that looks as though it has evolved through time, and so has generated a history, or a *past*).

Barbour's approach is certainly timeless in that it contains no background temporal metric in either the classical or quantum theory: the metric is defined by the dynamics. His view looks a little like presentism – the view that only presently existing things actually exist – but clearly differs since there are a bunch of other Nows that exist, only not in the same spacetime (but 'timelessly' in 'superspace' instead). This space of Nows might in fact be employed to ground a *reduction* of time in the same way David Lewis' 'plurality of worlds' provides a reduction of modal notions.[26] The space of Nows is given once and for all and does not alter, nor does the quantum state function defined over this space, and therefore the probability distribution is fixed too. But just as modality lives on in the structure of Lewis' plurality, so time can be taken to live on in the structure of Platonia.

The jury is still out on the coherence of Barbour's vision (philosophical and otherwise), but it makes a fun and worthwhile case study for philosophers of physics who like to think about time, change, and persistence. Not quite as radical as Barbour's are those timeless views that accept the fundamental timelessness of general relativity and quantum gravity that follows from the gauge-invariant conception of observables, but attempt to introduce a thin notion of time and change into this picture – these fall uncomfortably between the Heraclitean and Parmenidean notions.

A standard approach along these lines is to account for time and change in terms of time-independent correlations between gauge-dependent quantities. The idea is that one never measures a gauge-dependent quantity, such as position of a particle; rather, one measures 'position at a time,' where the time is defined by some *physical* clock. Carlo Rovelli's 'evolving constants of motion' proposal is made within the framework of a gauge-invariant interpretation. As the oxymoronic name suggests, Rovelli accepts the conclusion that quantum gravity describes a fundamentally timeless reality (so that there are only unchanging, time-invariant, constants of motion), but argues that sense can be made of change within such a framework by 'stringing them together.' Take as a simplified example of an observable m = 'the mass of the rocket.' As it stands, this cannot be a genuine (gauge-invariant) observable of the theory since it changes over (coordinate) time (it does not take on the same value on each time slice since it will be losing fuel as it travels, for example). Rovelli's idea is to construct a one-parameter family of observables (constants of the motion) that can represent the sorts of changing magnitudes we observe. Again simplifying, instead of speaking of, e.g. 'the mass of the rocket' or 'the mass of the rocket at t,' which are both gauge-dependent quantities (unless t is physically defined: i.e. the hand of an actual clock), we should speak of 'the mass of the rocket when it entered the asteroid belt,' $m(0)$, 'the mass of the rocket when it reached Mars,' $m(1)$, 'the mass of the rocket when it reached the human colony,' $m(2)$, and so on up until $m(n)$. These quantities are each gauge invariant (taking the same value in any system of coordinates), and, hence, are constants of the motion. But, by stitching them together in the right way, we can explain the *appearance* of change in a property of the rocket (understood as a persisting individual). The obstacles to making sense of this proposal are primarily technical, but philosophically we can probe the fate of persisting individuals given such a view, among other issues of perennial interest to philosophers.

A large portion of the philosophical debate on the problem of time tends to view it as a result of eradicating indeterminism via the 'quotienting procedure' for dealing with gauge freedom – as mentioned in the previous section, and as found in the hole argument literature: treating the physical structure as encoded in a reduced space involving as points the equivalence classes of gauge-related states. This point of view can be seen quite clearly in a 2002 debate between John Earman and Tim Maudlin,[27] where both authors see the restoration of determinism via hole argument type considerations as playing a vital role in generating the problem – but whereas Earman sees the problem as something that needs to be accepted and explained, Maudlin views the problem as an indication of a pathology in the (Hamiltonian) formalism that generated it.

The idea is that when the quotienting process is performed in a theory in which motion corresponds to a gauge transformation, time is thereby eliminated. But the world is frozen regardless so long as we deal with physical observables (those that are insensitive to the gauge-changes brought about by the constraints, and so the Hamiltonian). Hence, we could forget the quotienting, and let time merrily advance, but nothing observable will register this difference.

Earman and Maudlin couch their debate in terms of an argument against the existence of time due John McTaggart. In a classic 1908 paper, "The Unreality of Time," McTaggart argued that we can think of positions in time in two ways: first, each position is *earlier* than some and *later* than others (known as the *B-series*); and secondly, each position is either *past, present*, or *future* relative to the present moment (known as the *A-series*). The first involves a *permanent* ordering of events, while the second does not: if an event E_1 is ever earlier than event E_2, it is always earlier. But an event, which is now present, *was* future, and *will* be past. We met this kind of alteration in the ontological status of events (under the umbrella of 'becoming') in §4.3. This corresponds to the 'flow of time' (common to our experience of the world), consisting of later events becoming present and then past: time passes in this way. McTaggart had argued that both series are needed to account for time. To make sense of flow, we can think either of sliding the B-series 'backwards' over a fixed A-series or sliding the A-series 'forwards' over a fixed B-series.

The B-series is seen to depend on the A-series since the only way that events can change is with respect to their A-properties, not their B-relations (which are eternally fixed): the event 'the death of Alan Turing' does not change in and of itself, but, given the A-series, it changes by becoming 'ever more past,' having *been* future, relative to the Now. So 'Alan Turing is dead' changes its truth-value: during all of those B-series positions in which Turing is alive (or not yet born) the proposition is not true, but is true thereafter. But B-series versions do not have this time-varying feature: 'Alan Turing's death is earlier than this judgment' is either always true or always false. But this ordering isn't enough for time since there is no change.

So McTaggart concludes that time needs an A-series to ground the notion of change. But then he argues that the A-series, and therefore time, is contradictory. A-properties ('is past,' 'is present,' and 'is future') are mutually exclusive (they conflict), but events must posses *all* of them: a single event like 'Turing's death' *is* present, *will be* past, and *has been* future. The problem looks like it can be evaded by pointing out that no event has all three properties 'at the same time' (simultaneously), but this leads to a regress: *is* present, *will be* past, and *has been* future all involve some additional 'meta-moments,' which will also face the problem of being all of past, present, and future – the obvious save now is to invoke 'meta-meta moments,' and we are on the road to an infinity of such meta-levels. So McTaggart claims that time does not exist: the B-series can't cope with the demands of time on its own and the A-series, which it requires, is incoherent.

So much for the A- and B-series: what is left to put in their place? McTaggart suggests that an ordering of events remains, a *C-series*, but this cannot be temporal, for it does not involve change, being a serial ordering of events themselves – that we have a string of events, E_1, E_2, E_3, implies that there is any change no more than the ordering of the letters of the alphabet implies change. Modern philosophers of time are divided over whether an A-series (or something similar) in needed to make sense of time and change: the nay-sayers are grouped into the category of B-theorists or 'detensers' and the yea-sayers are grouped into the category of A-theorists or 'tensers.' The A-theorists will say that the B-theorists cannot properly accommodate the notion of the passage of time and can, at best, allow that it is an illusion. The B-theorist denies that 'passage' is necessary for time and change, and is happy to see it done away with. Both sides claim support from physics: B-theorists generally wield spacetime theories (such as special relativity) and the A-theorists wield mechanical theories such as quantum mechanics.

What is interesting about the problem of time in quantum gravity is that neither the A- nor the B-series seems to make sense anymore. To link up these old-fashioned philosophical concepts to quantum gravity, Earman introduces a character called 'Modern McTaggart,' who attempts to revive the conclusions of old McTaggart by utilizing a gauge-theoretic interpretation of general relativity. Earman is dismissive of A-theories, which he claims are not part of "the scientific image." But it is important to note that the debate here is not directly connected to the old debate in the philosophy of time between 'A-theorists' and 'B-theorists' (or 'tensers' and 'detensers'). Both of these latter camps agree that time exists in a sense (they are committed to some t in the physical world), but disagree as to its nature. By contrast, the division between Parmenidean and Heraclitean interpretations concerns whether or not time (at a fundamental level) exists, period!

Earman's response is to argue that general relativity is nonetheless compatible with change, though in neither the B- nor the A-series (nor the C-series) senses. Instead he introduces a 'D-series' ontology consisting of a time-ordered sequence of 'occurrences' or 'events' in which different occurrences or events simply occupy different positions in the series. This sounds like the C-series, but the occurrences here are the gauge-invariant quantities involving relationships between physical degrees of freedom (like Rovelli's evolving constants of motion, which Earman's strategy gives a philosophical expression of). There is no flow here, but there is change embedded in the different events laid out in different D-series positions. The only difference between the C-series and the D-series then is in the nature of the events that are strung together. In the case of a D-series picture one of the elements of a gauge-invariant quantity can be a physi-cal clock (a wristwatch), against which is considered some other element (such as your position relative to your front door) so that we can see how something mirroring experience might be elicited from this picture. But strictly speaking, nothing is changing here.

For Maudlin, any such interpretation is absurd. The absence of change should be a reason to reject whatever framework led to that con-clusion. Maudlin sees the problem to be taking a 'surface reading' of general relativity too literally, where such a reading cannot explain why a bizarre frozen-world conclusion is emerging. But, as we have seen, it can explain the changelessness and the appearance of change. The answer is related to the gauge-invariant response to the hole argument: change with respect to the manifold is ruled out; if we focus on those quantities that are independent of the manifold we can restore change by considering the 'evolving' relationships between these quantities. That is: change is to be found in things other than the manifold, namely in the relationships between physical degrees of freedom. That is, the crazy sounding results only come about by considering the wrong types of observable: gauge-dependent, unphysical ones. But this is to adopt a substantive response that buys into the gauge-theoretical interpretation!

Finally, if we agree with the Parmenideans that we don't have space and time at the level of quantum gravity, then where does the classical spacetime we appear to be immersed in come from? This leads to the problem of 'spacetime emergence.' In other words, saying that quantum gravity is timeless (and spaceless) leaves a challenge in squaring it with the low-energy physics that appears to involve such entities – not to mention the evidence of our senses. This is a difficult problem, but steps have been taken in doing exactly this.

In the author's view, quantum gravity offers the biggest open problem for philosophers of physics and it really ought to be a bread and butter topic: it is, with its novel views on time, change, space, persistence, and so on, a playground for philosophers.

What to read next

1 Craig Callender and Nick Huggett, eds. (2001) *Physics Meets Philosophy at the Planck Scale.* Cambridge University Press.
2 Jeremy Butterfield and Chris Isham (1999) On the Emergence of Time in Quantum Gravity. In J. Butterfield, ed., *The Arguments of Time* (pp. 111–168). Oxford University Press.
3 Dean Rickles (2008) Quantum Gravity: A Primer for Philosophers. In D. Rickles, ed., *The Ashgate Companion to Contemporary Philosophy of Physics* (pp. 262–382). Ashgate.

Research projects

1 Consider which (if any) philosophical account of persistence (of individuals) fits with Rovelli's evolving constants of motion approach.
2 Read Julian Barbour's book *The End of Time* (Oxford University Press, 2001) and decide whether his theory is *really* timeless.
3 Figure out whose side you're on (if any) in the debate between Tim Maudlin and John Earman over the use and interpretation of the constrained Hamiltonian formalism in general relativity.

8.6 Fact, Fiction, and Finance

Physics is powerful. It is often seen to be a *universal science*, providing the ontological roots of all of the other sciences. But just *how* powerful? Where does the domain of physics end and, say, the social sciences or psychology begin? Of course, most of us wouldn't hesitate in agreeing that any social system and human mind are essentially 'built from' components obeying physical laws: if they violated the laws of physics then so much the worse for those systems and minds. These days at least, the 'physics versus everything else' separation is attributed to the *complexity* of the systems, sometimes with a nod to a future in which physics can crack the problems of society and consciousness too: such features (including 'higher-level laws') will be 'emergent' properties of the physical systems realizing them.[28]

If we agree that economic systems are complex systems, then a natural policy for a physicist is to try to adapt the techniques of statistical physics (the complex system theory *par excellence*) to the description and prediction of economic phenomena. 'Econophysics' then describes a cluster of methods and models designed to capture the (statistical) properties and behavior of economic systems (usually just restricted to financial markets) using theories of physics (primarily statistical physics).[29] It sits, in terms of disciplines, more in physics than economics (a social science), and one can now find econophysics papers appearing in physics journals, side by side with papers on dark matter and quark-gluon plasmas.

But the simple 'econophysics = statistical physics of finance' link is an approximation. We can in fact discern several layers to this proposed physics-economics link:

- Econophysics involves a kind of *statistical mechanics* of financial markets.
- Econophysics is a proper *empirical study* of financial data: a data-first approach (rather than model-first), in which economic assumptions are not made at the outset.
- Econophysics involves viewing financial/economic systems as *complex systems* with interacting economic agents as parts.
- Econophysics is the general *application* of the models, theories, and concepts of physics (not just statistical physics) to financial markets.
- Econophysics is the idea that financial markets (and economic systems more generally) follow similar *laws* (or the same *kind* of laws) as 'natural' systems.
- Econophysics involves finding *analogies* between financial/socio-economic and physical (natural) systems.

The motivation behind each separate conception is the potential for prediction (and control) and explanation of some market quantities (price, volume, or volatility of some 'financial instrument') or events (notably, bubbles and crashes – i.e. large fluctuations or 'extreme' events – and more general market trends), through the formulation either of equations of motion derived, in this case, through the recognition of patterns and signatures in the data or the formulation of statistical laws that the quantities or events obey. How does this look in practice? At the core is a postulated mapping between a general model from physics (a model of cooperative phenomena) and the behaviors of economic agents – note that "cooperative" in this context does not mean the nice, social idea of working together for the greater good, but simply the degree of *imitation* involved among a system's parts.

The global financial crisis that began around 2007 is commonly believed to be traceable to the relative ease with which borrowing for home loans could be done before that time – fueled by the widespread belief that property was as 'safe as houses' and so allowing for the creation of 'subprime mortgages' given to those with low credit ratings (i.e. not enough money to handle dips in the value of whatever the loan was for, or rises in external living values). But the banks weren't stupid (just highly unethical!): they charged more (in interest) to those able to afford less, in order to cover their own increased risk. This led to a boom in house buying, with prices increasing and the expectation that they would continue to increase – some used the value stored in their existing homes to fund other homes. But such a situation is naturally very fragile to 'shocks.' The complexity features come from this aspect: there is an interdependency between the various economic agents in that they require that the prices keep going up (with low interest rates) so that they can keep up with their payments – seeing buying occurring brings in yet more buyers (driving the prices up) troubled by the 'fear of missing out' (FOMO: which results in a kind of human-herding behavior). This naturally leads to a greater chance of slipping behind in mortgage payments and ultimately 'defaulting' on their loan (losing their home) if anything should alter the rising price trajectory – even small dips could result in significant losses of property for those with subprime mortgages. As soon as the foreclosures start, supply increases, and prices reduce, causing more selling in a bid to avoid large losses, which triggers more selling at a faster rate (driving prices down) and so on in a snowball effect until the financial equivalent of an avalanche occurs.

Hence, we have the economic notions of boom and bust, or bubbles and crashes. This can be linked quite naturally to aspects of cooperative phenomena in statistical physics. The trick involves viewing economic agents along the lines of a 'spin model' of ferromagnetism. Each agent has a set of states that they can occupy, switching between them according to certain intrinsic properties (e.g. such as how risk-averse they are), but also, crucially, able to become dependent on the structural properties of the economic system as a whole by 'nearest neighbor interactions,' which can spread over a system in certain contexts. Hence, just as magnetism occurs when the spins of atoms in the material point in the same way (demagnetizing otherwise, when they point in random directions), so economic agents can 'point the same way' (all buying or all selling at the same time, thus generating bubbles and crashes). In the case of magnetism one needs a parameter, such as temperature, that can be tuned to generate the cooperative behavior of the atoms: heating causes randomness, but cooling causes an orderly structure to emerge.

In between these two phases (order and disorder) is a 'critical point' relative to the tuning of the parameter, characterized by very specific statistical properties involving events of all sizes distributed according to a power law (roughly: many small events, not so many mid-sized events, and very few large events, as with the distribution of earthquake magnitudes). Such distributions signify complexity in the system, with parts heavily interdependent.

Economic bubbles are then modeled as systems in a critical phase. The cooperative tendency here is simply the herding instinct as above and is modeled by long-range correlations that spread over the system. In such critical, highly cooperative scenarios even tiny shocks to the system can trigger calamitous effects, such as crashes (i.e. spontaneous switches to some value). In these cases it doesn't make sense to say that some individual event *caused* the crash: the system was led into an unstable state over time by internal processes. Crashes are systemic (endogenous) making 'causal postmortems' very difficult.

The most common model of this type is the Ising model. This models the cooperative coupling between elements by a parameter K, and the tendency to disorder by a noise term (or rather its amplitude) σ. There will be order if K dominates and disorder if σ dominates. Crucial is the existence of a 'critical value' K_c of K, which separates these phases: when $K < K_c$ there is disorder and little cooperation or herding (because the elements are effectively disconnected there is no threat of bubble-type behavior emerging). As K tends toward K_c clusters of order form, in which the elements are aligning their values in such a way that alignment in one region can spread, causing alignment in other areas, meaning that the elements are now interconnected. This means that a small shock, which would otherwise be insignificant, can cause an entire system to shift. At the critical point one finds that the system is 'scale invariant' in the sense that zooming in and out one finds the same statistical features (relative to the temporal or spatial resolution): it is a fractal. The interconnectedness is, according to econophysics, behind the crashes: they are collective effects or emergent features. Poised at a critical point K_c (in which there is symmetry in the possible states), small influences can break the symmetry (corresponding to $K > K_c$) and self-organization occurs.

There are other aspects that suggest a statistical physics approach, and that are similarly described. In financial economics there is a group of features known as 'stylized facts,' which are properties that are common across many financial objects (stocks, bonds, etc.), markets (money, gold, property etc.), and time periods. For example, prices and their changes appear to be random; returns appear not to be random since their statistical distributions possess so-called 'fat tails' associated with the power-laws mentioned above (showing a greater likelihood for extreme values than a simple random, Gaussian distribution would allow); and finally, volatility (fluctuations) is not uniformly distributed, but clusters so that there are highly volatile and non-volatile periods (big price changes, of either sign, follow big price changes and little ones, of either sign, follow little ones). This suggests that the standard model of economics (based on the randomness assumption) is not mapping onto economic reality: stock market crashes of the magnitude of, say, the 1929 Wall Street or 1987 ('Black Monday') crashes, should not be occurring as often as they do. Econophysicists claim to be closer to reality in this respect.

Very often in systems with interacting parts, and whose interacting parts generate properties of the unit system, one finds that the thus generated properties obey scaling laws. Scaling laws tell us about statistical relationships in a system that are invariant with respect to transformations of scale. In statistical physics these scaling laws are viewed as emergent properties generated by the interactions of the microscopic subunits. Scaling laws are explained, then, via collective behavior among a large number of mutually interacting components. The components in this financial case would simply be the market's 'agents' (traders, speculators, hedgers, etc.). These laws are 'universal laws,' independent of microscopic details, and dependent on just a few macroscopic parameters (e.g. symmetries and spatial dimensions). Econophysicists surmise that since economic systems consist of large numbers of interacting parts too, perhaps scaling theory can be applied to financial markets; perhaps the stylized facts can be represented by the universal laws arising in scaling theory. This analogy is the motivation behind a considerable chunk of work in econophysics; it is through this analogy, then, that the stylized facts receive their explanation – though presumably not their 'ultimate explanation,' which will involve such things as the agents' psychology, the institutions in which the agents operate, and so on.

This 'scaling' (or scale invariance) is, then, at the root of the transference of statistical physics to finance. Notably, from the point of view of complexity research, the power law distributions are scale invariant (like fractals): events (or phenomena) of all magnitudes can occur, with *no characteristic scale*. What this means is that the (relative) probability of observing an event of magnitude $|x| = 1,000$ and observing one of $|x'| = 100$ does not depend on the standard of measurement (i.e. on the reference units). The ratio between these probabilities will be the same as that for $|x| = 1,000$ and $|x''| = 10,000$. Hence, there is no fundamental difference between extreme events and events of small magnitude: they are described by the same law. Specifically: near a critical point, fluctuations of the (macroscopic) order parameter will appear at all possible scales. In the case of the more common liquid-gas phase transition one will have liquid drops and gas bubbles ranging from the molecular level to the volume of the entire system. Hence, at the critical point these fluctuations become (theoretically) infinite. The analogous situation in the financial context would be, for example, fluctuations in asset returns at all possible scales.[30]

What distinguishes econophysics from other approaches in economics is the interpretation of the stylized facts. In addition to viewing them as emergent properties of a complex system, it also includes a greater *commitment* to the stylized facts, treating them not only as a central guide to the nature of economic reality, but also as genuine *laws* rather than 'mere' local regularities. Part and parcel of this view of the stylized facts as genuine laws is that it simply doesn't matter whether we view the variables of a statistical physics model as spins of atoms or economic agents since the laws themselves are emergent in precisely the sense that they don't depend on the microscopic details of the system.

However, this stance is, on the face of it, rather hard to square with the addition of decision-making agents (with *free will*): such agents are surely incompatible with the possibility of invariances, and without invariances there are no symmetries, and without symmetries there are no laws. We might think that, on this basis, like economics, econophysics can at best be a descriptive historical science, analyzing what already happened in some *particular* economic situation. Other econophysicists (most, in fact) believe that they can find some laws for market dynamics, albeit statistical ones, of course. But, it might be objected, whereas laws of nature are independent of initial conditions and don't make reference to specific, particular systems, socioeconomic laws seem not to be of this sort, varying from country to country and institution to institution: markets just embody what economic agents do, and since these agents have free will, they can change at a whim. This could affect the more general statistical laws too since the distributions are determined by the collective behavior of the economic agents. But the econophysics argument insists that these collective properties and laws are sufficiently robust that they can ignore the human nature of the agents.[31] After all, if humans are not viewed as supernatural 'magical beings' then they must surely be governed by physical law. The proof is no doubt in the pudding, and we must await the construction of statistical models that enable reliable predictions about (distributions of) financial properties and their fluctuations. Moreover, a thorough analysis must proceed by carefully unpacking exactly what is meant by 'law,' 'free will,' and the other central (and difficult) concepts relevant to the debate – in other words: lots of interesting work for philosophy!

What to read next

1 Didier Sornette (2003) *Why Stock Markets Crash: Critical Events in Complex Financial Systems*. Princeton University Press.

2 Neil Johnson et al. (2003) *Financial Market Complexity: What Physicists can tell us about Market Behavior*. Oxford University Press.

3 Dean Rickles (2011) Econophysics and the Complexity Of Financial Markets. In C. Hooker, ed., *Handbook of the Philosophy of Science. Volume 10: Philosophy of Complex Systems* (pp. 531–565). Elsevier.

Research projects

1 Can econophysical models really be said to *explain* economic phenomena?
2 Are 'stylized facts' examples of emergent properties?
3 How do we get around the problem of 'free will' in econophysics' modeling? What becomes of invariances?

8.7 Anthrobatics and the Multiverse

Anthropic reasoning, as the name suggests, involves using the existence of human beings (or rather *observers*) as 'data' from which to make (usually rather deep) inferences about the nature of reality and most often when other data is hard to come by (for example, in the case of why the fundamental constants have their values or why we live in a space of three dimensions). The world must be compatible with our existence as observers, so our existence might furnish us with a useful scientific tool.

It is widely believed that Anthropic reasoning is bad 'unscientific' reasoning. A passage from Douglas Adams makes the shortfalls clear:

> Imagine a puddle waking up one morning and thinking, 'This is an interesting world I find myself in, an interesting hole I find myself in, fits me rather neatly, doesn't it? In fact, it fits me staggeringly well, must have been made to have me in it!' This is such a powerful idea that as the sun rises in the sky and the air heats up and as, gradually, the puddle gets smaller and smaller, it's still frantically hanging on to the notion that everything's going to be all right, because this World was meant to have him in it, was built to have him in it. (D. Adams (2002) *The Salmon of Doubt: Hitchhiking the Galaxy One Last Time*. Harmony Books, p. 131.)

Such 'Puddlethropics' makes very clear the ludicrous nature of *a certain kind* of Anthropic reasoning, involving the 'fine tuning' of parameters (here the shape and depth of a hole). Many aspects of the world look 'purpose-built' for human existence, leading some to believe, like the poor puddle, that the world is made for them.

The standard example of Anthropic reasoning is that going against Kepler's theory of the Earth's orbit. Kepler believed that a fundamental principle should pick out uniquely the orbits of the planets around the Sun, and he based this in a system of nested Platonic solids. The Anthropic explanation simply points out that if the Earth were *not* at this orbit, the chemical composition of the planet wouldn't have given rise to creatures capable of making observations about the orbits! Hence, here the mystery is taken out of some phenomenon. But we need to be careful not to lapse from this uncontroversial claim into puddlethropics. We might rightly feel shortchanged with the explanation as it stands: surely there is still a deeper explanation based on fundamental laws or something beyond? Pointing to our existence doesn't really explain why the orbits are as they are at all. We want to derive the orbits from some initial conditions and laws or some 'principle.'

But there is a different kind of Anthropic approach (or an additional component to the above Anthropic explanation), which also removes some of the mystery: we can point out that the Earth is one of countless billions of planets, some of which will naturally be in the right configuration relative to their stars to support life. Our situation might look finely tuned, but it is simply what we should expect to find given our constitution. Any other worlds containing observers will have the same conditions, and worlds with these conditions form a small segment of possible conditions in a larger space: nothing is 'tuned' here at all. The problem with the other form of Anthropic explanation assumed some kind of *uniqueness* of the Earth's position, which makes it seem like the Earth's convenient position is for our benefit (thus pointing to Flying Spaghetti Monsters directing the show!). Remove this and we remove the tuning idea, and along with it many standard objections. However, there is no known theory that can do this. String theory once hoped to achieve this kind of unique prediction of apparently finely tuned parameter values, but was found to generate an ensemble of solutions instead (known as 'the Landscape').[32] But one must be careful in using ensemble-based forms: the ensemble must be known to exist or be well-motivated, otherwise it will have an ad hoc, *Deus ex machina* character.[33]

As with possibilities, then, there are several grades of Anthropic principle and associated reasoning, linking to this arbitrary versus motivated usage of ensembles, to the interpretation of fine-tuning, and to the way the Anthropic data is used: the 'weak' [WAP] and 'strong' [SAP] are the two most important:

WAP: this corresponds to the mystery-reducing response to Kepler's orbit problem above. Brandon Carter (who coined the term 'Anthropic principle') states it as: "What we can expect to observe must be restricted by the conditions necessary for our presence as observers" ([4], p. 291). This seems to have the status (almost) of a tautology: *of course* a necessary condition for X must be in place for X! 'Almost' because it is really pointing out a selection-bias that can infect our sci-

entific inferences, namely that what we see is much the same as what
there is *everywhere and everywhen.*

SAP: the strong principle essentially generalizes WAP by imposing it as a
universe-wide Copernican principle (rather than WAP, which invokes
a kind of localized version): we are typical of the universe, rather than
just a small patch – it involves the idea that the universe *must* be such
so as to allow our existence as observers. It is the generality of the
inferences made that are much stronger. To make sense of SAP an
ensemble of universes is required, in which our universe is typical.

In either case, to be scientifically useful, the reasoning needs to make it that
our scientific observations are made *more likely* given the addition of the
Anthropic premise. If we have an ensemble of worlds at our disposal then
sense can be made of this in simple relative-frequency terms: most observers will be like us in the ensemble (or in the restricted subset allowing
observers); some will not, and will have very different conditions (maybe
existing in wormholes, six-dimensional spacetime, or who knows what).
Those observers need to be unlikely, since we don't see worm-holes about
us, and we seem to be in four-dimensional spacetime. If it were found that
observers were far more likely to spring up from wormholes than our kind
of universe, then we would have a bad explanation of our world since it
would be hard to see ourselves coming out as a conclusion. Any theory we
give about the makeup of the world must make us typical. That is the key
point.

As mentioned, the standard alternative to Anthropic-style explanations
is law-based explanations, where the laws, interactions (forces of nature),
and constants of nature are invoked instead. But what if we want to explain
these very laws, interactions, and constants? We can't very well invoke
them to explain themselves without circular reasoning. This is, then, a
common problem to which Anthropic reasoning is applied: explaining
the seemingly arbitrary values of the fundamental parameters featuring
in our laws of nature. If we were to plot in an abstract space, all possible
parameter values, we would find that only a tiny region of this space would
possess values fit for observers like us. But in order to make Anthropic reasoning work, you need a lot of worlds. Enough worlds to make ours a *likely*
prospect. Why might we think there are multiple universes? Aside from the
many-worlds interpretation the strongest reasons come from string theory
and inflationary cosmology – the latter suggests that our Big Bang was not
a unique event. The ensemble needs to be highly variegated too, so that
once again ours isn't privileged in some way. This also gets around a hidden
assumption (buried in any 'explanation-seeking *why* question'), which we
have seen in the Leibniz shift arguments, concerning contrasting possibilities: why is the world *here* rather than *there* (in absolute space). In order to

ask such a question, then, there are usually alternatives. Having a varied ensemble gives us a contrast class: why *these* laws rather than *those*.

A more serious problem, in terms of starting from our existence and figuring out the conditions *necessary* for this existence is: how do we know what those conditions are? Any uncertainty here trickles down into uncertainty about inferences made from them. That is, we need to know what kind of observers we are, what would need to be in place in the world to allow for observers of that kind, and only then can deductions be made about what we can *expect* to observe (with those conditions hopefully increasing those expectations in cases of puzzling phenomena such as the three-dimensionality of space). To know what conditions are required for our presence as observers, we need to know what kinds of observers we are, then. That's not a trivial matter. Is carbon a necessity? If so, then certain cosmic conditions must be in place, putting observers at a certain phase of the universe's evolution. So our temporal position is an important consideration. So the observations we make are not typical in an unrestricted sense, for the universe as a whole, because we are not typical in this sense. But, the Anthropic reasoner will say, what we observe *is* typical for *observers* because observers require what is observed for their very existence.[34] This is precisely what links Anthropic reasoning to the issue of 'fine-tuning': the finely tuned values will also be those that are typical in that subset of worlds fit for observers.

But again: what is an observer? Or in other words, fine-tuning for *what* exactly? It should be clear that 'Anthropic' is something of a misnomer: the observer-principle would perhaps be a better name, since the same reasoning applies to anything capable of making observations, whether from Earth or from the nearest habitable star system. But here the problems take better focus, since depending on what these observers are like, we will find different conditions necessary to support their existence.[35] The Anthropic principle lets us impose restrictions on the number of ways the world could be that would otherwise be hard to come by: it is a possibility carving tool. Any physics that conflicts with our presence can be put on the scrapheap and our presence should be expected given whatever physics we come up with.

A common objection to the Anthropic multiverse approach to explanation of the nature of things is that it trivializes the answers we give. It's rather like the person that answers some question about 'which state of the USA is the largest' by listing all 50: they'll be right at some point! Burton Richter (who directed SLAC) calls the Anthropic multiverse a "metaphysical wonderland" ([40], p. 8) and complains that "much of what currently passes as the most advanced theory looks to be more theological speculation." But it's clear however, that Richter misunderstands what is really going on when he says:

The Anthropic principle is an observation, not an explanation. To believe otherwise is to believe that our emergence at a late date in the universe is what forced the constants to be set as they are at the beginning. If you believe that, you are a creationist. We talk about the Big Bang, string theory, the number of dimensions of spacetime, dark energy, and more. All the Anthropic principle says about those ideas is that as you make your theories you had better make sure that α [the fine structure constant] can come out to be $1/137$; that constraint has to be obeyed to allow theory to agree with experiment. I have a very hard time accepting the fact that some of our distinguished theorists do not understand the difference between observation and explanation, but it seems to be so. ([40], p. 9)

This presupposes that we don't have independent reasons for believing in an ensemble. With an ensemble to hand, we can focus in on the worlds with $\alpha = 1/137$ and try to show that it is expected when conditioned on our existence as observers. But he does point to a problem, which is that showing that a probability distribution is peaked on our world (and not on worlds with other values) is no simple matter: how do we calculate probabilities of these worlds? In terms of direct experience, we only have one universe to play with, so it isn't like pulling balls out of an urn or throwing dice. A principle of 'mediocrity' (or the Copernican principle) is imposed by default as one way to handle probabilities in this context; it says that no world in the ensemble is inherently more probable than any other.

Part of the problem is that there is an underdetermination of theory by data here, much like the one facing the flatlanders in §5.1, only here between ensemble and non-ensemble interpretations: the ensemble of worlds is not observable, so we can only make inferences about it. If we aren't just adding ghosts by adding an ensemble, then we need to see what it can do (as with Newton's absolute space and time required for making sense of inertia).

Another part of the root of the debate is a different mindset as regards what counts as a good explanation: a unique one, so that our world drops out as the sole solution (a 'theory of everything'), versus a non-unique one, in which our world is part of a space of solutions. This translates into thinking that our world is privileged or special versus unprivileged and non-special. Curiously, critics of Anthropic reasoning often claim that Anthropic reasoning violates the spirit of the Copernican revolution: it removes the *specialness* of our planet's place in the world (as the center of everything). Yet in wishing for uniqueness they make the same mistake, only at the level of the universe rather than the Earth.

What to read next

1   John Peacock and Alasdair Richmond (2014) The Anthropic Principle and Multiverse Cosmology. In M. Massimi, ed., *Philosophy and the Sciences for Everyone* (pp. 52–66). Routledge.
2   Nick Bostrom (2002) *Anthropic Bias: Observation Selection Effects in Science and Philosophy*. Routledge.
3   Bernard Carr, ed. (2007) *Universe or Multiverse*? Cambridge University Press.

Research projects

1   Actively seek everyday examples of Anthropic-type reasoning to build your capacity for identifying them. In each case, figure out why Anthropic reasoning (rather than a more orthodox form) is being used?
2   Does the fact the *several* different theories in modern physics involve an ensemble of other worlds better the hopes of Anthropic explanations of deep facts about *our* world?
3   Is it more *reasonable* to think that our world (including its laws and parameter values) might be explained by some unique equations or by showing how it is one among many other worlds?

Glossary

There are many central concepts in this book that will be unfamiliar to most readers. Here we provide simplified definitions/explanations of some likely candidates. These are not intended to be complete or fully accurate, and there are in almost every entry complications and qualifications.

Copenhagen interpretation: an interpretation of quantum mechanics built on 'the principle of complementarity,' namely the idea that quantum objects exhibit a duality according to which they can sometimes behave like a particle and sometimes like a wave, but never both simultaneously. Whether or not we measure the system determines which way it behaves, with the continuous wave 'collapsing' upon measurement to some discrete value. The particle and wave aspects are related by the uncertainty principle whereby measurement of the wave aspects (e.g. momentum) renders position uncertain.

Decoherence: the suppression of quantum interference effects through the interaction of a measured quantum system first with the measurer and then with the environment (or the universe) in which both the measured and measuring system are embedded.

Degree of freedom: the number of degrees of freedom correspond to the number of variables required to uniquely specify a state. In classical physics, a single particle has six degrees of freedom (three position and three momentum coordinates).

Diffeomorphism invariance: a diffeomorphism is a transformation that maps one manifold onto another by mapping points to one another. Such manifolds and their points are used to represent spacetime and its spacetime point-events. A diffeomorphism can be used to 'drag around' mathematical structures that represent physical objects (e.g. fields and particles) living at such points. General relativity is invariant with respect to such transformations in the sense that the laws of the theory are insensitive as to whether one has been performed: shifting everything by a diffeomorphism does not alter the observable physical state of the world.

Dynamics: the behavior of a system under the action of forces and constraints.

Entropy: a measure of order/disorder nowadays understood probabilistically in terms of the number of ways some configuration can be realized – the fewer ways to realize a configuration (i.e. the less likely it is), the lower the entropy.

Eigenstates and eigenvalues: in quantum mechanics observables O are represented by matrices, the eigenvalues λ (of the matrix) represent the measured value of O when the system is in a state ψ and when the following equation (the 'eigenvalue-eigenstate' link) holds: $O\psi = \lambda\psi$ (in which case, ψ is said to be an eigenstate).

Epistemology: at its most general, epistemology is the study of knowledge and what constitutes *true belief*. We are primarily concerned with scientific knowledge and the extent to which our physical theories allow us to gain knowledge about the world (along with the shape of that knowledge).

Equilibrium: a state characterized by the invariance (i.e. constancy) of some important feature of a system.

Equivalence class: the result of dividing up a collection of objects according to some 'sameness' relation (e.g. same shape, size, etc.). The members of an equivalence class are 'the same' in the sense that matters, so that any element will do to represent the entire class.

Equivalence principle: the statement that gravitational and inertial forces (such as those accelerations that push you back in your seat during takeoff in an airplane) are indistinguishable from the point of view of local experiments (i.e. those made in small regions).

Euclidean space: simply, the space in which one can do Euclidean geometry. It is a flat (i.e. uncurved) space of this kind that is used in Newtonian physics.

Expectation value: the expectation value of a quantum state is a real number (i.e. that can correspond to some measured value) – this number corresponds to the average value that you would get after performing measurements on a very large quantity of identically prepared quantum systems.

Galilean invariance: the sameness of (non-relativistic) physical events with respect to a switch (a Galilean transformation) that turns one inertial frame into another (e.g. by reorienting, moving, waiting some amount of time, or changing the velocity).

Group theory: the branch of mathematics that lets us represent symmetry by showing how properties and relations of some set of objects (the group elements) stay the same with respect to some transformation on them.

Haecceity: also known as a 'primitive identity' or 'thisness' ('*haec*' is Latin for 'this'), a haecceity is a non-qualitative property that makes an object the unique thing that it is. If a pair of objects differ *only in that there are two* (i.e. they are qualitatively identical), then the two are said to differ solely haecceitistically. Haecceitism is the view that there can be possibilities that differ in this way (its denial is anti-haecceitism).

Hidden variables: deterministic (non-random) variables postulated to explain the curious probabilities of quantum mechanics. The variables are not observable by us, and so the world *appears* random.

Identity of indiscernibles: the principle, due to Leibniz, that there can be no two objects that differ solely in that there are two of them (i.e. that are identical in every way). In slogan form: there is no distinction without a difference.

Inertia: resistance of an object to accelerations of changes in direction of motion. The forces that are felt during such changes are known as inertial forces (or inertial effects). An inertial reference frame is one in which there are no such forces (i.e. motion is constant).

Initial conditions: these are details about a system's state that are fed into the equations of motion describing some system to yield its past, present, and future behaviors.

Interference: an overlapping of two or more waves caused by a difference in their phases. When the waves are in phase they reinforce one another (constructive interference); when they are out of phase they cancel one another out (destructive interference).

Isometry: an isometry is a mapping that preserves all lengths. Such mappings include rotations, translations, and reflections.

Isomorphism: a one-to-one correspondence between a pair of objects implying that they are structurally identical (preserving properties and relations). When the isomorphism relates an object to itself it is called an automorphism.

Kinematics: The physical description of motion in spacetime in the absence of forces and constraints (i.e. objects in free motion).

Manifold: a manifold is one that looks like good old flat Euclidean space 'close up' (locally, that is, or in a point's 'neighborhood') but looks different when one 'zooms out' (globally): the apparent flatness of the Earth in our vicinity as compared to the view of the Earth from space gives a good approximation of this idea. Manifolds can be patched together, smoothly, from such small flat parts.

Metric: a function (on spacetime) for determining the distance (and angle) between a pair of points (or vectors) in space, time, or spacetime. In pre-relativistic physics the distance in spacetime could only ever be positive. In relativistic physics the distance can be negative.

Model: a simplified representation of a system. In this book we are interested in mathematical models in which a correspondence is set up between a mathematical structure and some physical phenomenon of interest.

Observable: a measurable quantity in the context of physical theory (e.g. position, temperature, etc.). In classical (i.e. non-quantum) physics it is a real-valued function on phase space (the space of classical states). In quantum physics it is an operator on Hilbert space (the space of quantum states).

Ontology: the aim of ontology is to classify and catalogue reality (what entities there are, what properties they have, and so on) in such a way that anything that happens in the world can be given an explanation by referring to the entries in this catalogue (e.g. by pointing to some particles and laws that they obey).

Phase: a number (ordinarily between 0 and 360) describing the wave aspect of some system (specifically the 'distance,' in the cycle, to the wave's next peak or trough).

Phase space: the space of all possible states for a system such that each point represents a distinct possible assignment of values to the system's variables.

Planck's constant: the fundamental constant that describes the relationship between frequency and energy of a wave and, loosely, determines the degree of 'quantumness.' This can be seen most easily in the Planck–Einstein relation $E(\text{nergy}) = h \times \text{frequency}$. These days physicists use the 'reduced Planck's constant' (or Dirac constant), which divides Planck's constant by 2π, on account of the quantization of angular momentum (or spin).

Possible world: a 'way the world could be.' Possible worlds are objects that make true modal talk (having to do with necessity and possible). There are grades of possibility involved: physically or nomologically possible worlds are those worlds (e.g. objects in spacetime, with various properties, entering various relations) that are consistent with the laws of physics. Metaphysically and logically possible worlds must simply be either conceivable or else logically consistent.

Probability amplitude: the complex number whose modulus squared (or absolute square) yields the probability for some event to occur.

Proper time: the time on a clock (or wristwatch) that would move along at rest with some object or particle. This is distinct from coordinate time and relies purely on coordinate-independent, objective physical processes (such as the number of revolutions of the clock hands).

Quantization: the process of converting some classical theory to a quantum theory. The details involve promoting classical variables from the classical theory to operators that must satisfy certain relationships governed by Planck's constant.

Quantum operator: mathematical representation (involving a special kind of matrix) of an observable magnitude in quantum mechanics. Among other things, matrices are needed because of their 'many valuedness,' which matches the fact that quantum mechanics deals with probabilities for outcomes rather than definite outcomes.

Reference frame: a system for locating events in space and time using, e.g. rods and clocks.

Relationism: the belief that space, time, or spacetime are not fundamental entities in the world, but are instead somehow emerge from relations between objects. Usually defined as the denial of substantivalism.

Relativity principle: a statement that the laws of physics are invariant with respect to some (group of) transformations so that they stay the same in any reference frame related by such transformations (which then form the theory's symmetry group).

Schrödinger equation: equation describing the evolution of the wave-function in quantum mechanics (and so the evolution of the quantum state).

Simultaneity slice: (aka 'spacelike hypersurface') this provides a way of building an instant from the set of events that occur at the same time. In effect, space (extension) is used to define time. We can imagine then building up a spacetime from these spatial slices by stacking them up where the slices are said to be a 'foliation' of spacetime.

Singlet state: two particles (usually electrons) prepared together in such a way that they have zero total angular momentum (or spin) – "singlet" refers to the fact that there is only one such two-particle state with this feature of zero spin.

Spin: a kind of quantized angular momentum that characterizes some quantum systems. An electron can have only one of two values (up or down).

State: the information summarizing the instantaneous condition of an object. In classical physics the state is represented by a point in phase space. In quantum mechanics the state is represented by a wavefunction. The state is plugged into equations of motion to make predictions about observed outcomes (or probabilities of such in the case of quantum mechanics).

Substantivalism: the view that space, time, or spacetime exist *over and above* the objects contained within it. In other words, if you were to magically remove all matter and energy in the world, space would remain.

Superposition principle: in quantum mechanics this is the property that if there are a pair of possible wavefunctions then their combination is also a possible wavefunction, where interference between these two other states can be exhibited.

Symmetry: an operation (such as a rotation or reflection) that leaves some (or all) features of an object or structure unchanged. The group of transformations that leaves the object unchanged (or invariant) is known as the symmetry group.

Topology: a kind of generalization of geometry to 'deeper' properties of a space or surface (namely, those that remain unaffected by transformations that don't put holes in the space, including stretches, twists, etc.). This leads to more general ways of distinguishing, identifying, and classifying manifolds – famously, for example, a coffee cup is topologically the same as a doughnut (since it has the same number of holes).

Wavefunction: a mathematical representation of the state of a system, usually denoted by ψ, the absolute square of which corresponds to a probability to find a particular value for some observable (e.g. location of a particle in space). The wavefunction evolves according to Schrödinger's equation and lives in a kind of vector space known as Hilbert space.

Worldline: the path traced by an object (usually a particle) through space-time. Each point of the worldline represents the particle at a particular time and point of space. The shape of the worldline indicates the kind of motion the object undergoes: curved lines correspond to accelerated motion; straight lines to constant motion.

Notes

CHAPTER 1 INTERPRETING PHYSICAL THEORIES

1 There have been some interesting developments in the physics–philosophy 'dialogue' recently, showing that the question over the legitimacy of thinking philosophically about physics (and science in general) is still alive. Stephen Hawking declared (at Google's 'Zeitgeist' Conference in 2011) that "philosophy is dead" on account of its detachment from actual, current science: philosophers just haven't kept up with the science. A read through journals such as, e.g. *Studies in the History and Philosophy of Modern Physics*, will quickly reveal a very different story, with philosophers writing on subjects at the cutting edge (often in collaboration with physicists). A further problem with Hawking's view is that he ignores the fact that a great many philosophical assumptions are buried in the work he (and other theoretical physicists) do, especially in terms of how mathematical structure is understood to map onto the world. Likewise, in his book *A Universe from Nothing: Why there is Something Rather than Nothing* (Atria Books, 2013), physicist Lawrence Krauss argued, following Feynman's path, that philosophy is inert when it comes to physics: it doesn't influence how physics works, nor does it progress like physics. In answering (he thinks) the most fundamental problem of philosophy (why there is something rather than nothing) *using physics alone*, he doesn't see any elbow room left for philosophy to do its thing: even the deepest questions can be dealt with by physics. Philosopher of physics David Albert responded (in the *New York Times*, March 23, 2012) with a critique outlining how Krauss hadn't answered this deep question at all, but an entirely differerent one: how can you get something from not quite nothing (i.e. something)! For example, it leaves completely untouched the question of where the laws of physics come from, and requires a quantum vacuum (very much a *something*). Read Krauss' book, and see if you think he has answered the question satisfactorily.

2 I'm sure historians of science would quibble with much that I have said here, finding earlier examples of laws of nature of sorts (e.g. Aristotle's notion of 'natural place' guiding fire, water, and so on), or examples of early uses of instrumentation (e.g. the use of gnomons to chart the progression of the sun's path) – see Daniel Graham's *Science Before Socrates* (Oxford University Press, 2013) for just such a history. This might well be so, but the point I'm making is that the detachment of scientific knowledge from our *unaided* sense organs combined (or going hand in hand) with the increased mathematization of scientific knowledge provides fertile ground essential for doing modern philosophy of physics.

3 I point the reader toward Wesley Salmon's excellent collection *Zeno's Paradoxes* (Hackett Publishing Company, 1970) for more examples and discussions.

4 Readers interested in more details should consult Roman Frigg and Stephan

Hartmann's "Models in Science": http://plato.stanford.edu/entries/models-science/.

5 These are difficult issues, which we won't pursue further in this book. However, Christopher Pincock provides an admirable overview and attempt at an explanation of the role played by mathematics in science in his book *Mathematics and Scientific Representation* (Oxford University Press, 2012). This is especially useful from the point of view of Wigner's question since it also deals with *failures* of mathematical modeling in which there is too much distance between the model and reality (e.g. omitting or idealizing in such a way as to generate unphysical predictions in the theoretical description).

CHAPTER 3 SYMMETRIES IN PHYSICS

1 Any book on group theory will explain this basic idea – a good choice for those interested in physics applications is Chris Isham's *Lectures on Groups and Vector Spaces for Physicists* (World Scientific, 1989). The Galilean group is strictly speaking a 'Lie group' since the various parameters are continuous. A finite rotation, of a planet for example, would be *generated* (or 'built up') by the accumulation of lots of infinitesimal rotations. A useful book on Lie groups, for those already acquainted with vectors and matrices, is Harriet Pollatsek's, *Lie Groups: A Problem-Oriented Introduction via Matrix Groups* (Mathematical Association of America, 2009).

2 This is very over-simplified and there are many more subtle issues surrounding the proper interpretation of the symmetry group of general relativity, but these are too technical to go into here. The interested reader should consult §4 of C. Rovelli and M. Gaul, "Loop Quantum Gravity and the Meaning of Diffeomorphism Invariance" (in J. Kowalski-Glikman (ed.) *Towards Quantum Gravity*, Springer: pp. 277–324).

CHAPTER 4 GETTING PHILOSOPHY FROM SYMMETRY

1 See §5.1 of his *Space, Time, and Stuff* (Oxford University Press, 2012).

2 The details are a little complicated, but philosophically very interesting. The theory is, in a sense, timeless, making do with the instantaneous configurations and intrinsic differences between them. Space and time are given over to configuration space (usually viewed as an abstract framework for talking about things in space and time), though with the usual symmetries inherited from Newtonian space and time removed, so that each point in the configuration space represents all of the configurations of Newtonian space and time that are isometric. A relationalist theory would then take place relative to this space instead. I refer the reader to Julian Barbour's popular book on the subject, *The End of Time* (Oxford University Press, 2000).

3 This is very much a 'Leibniz for dummies' approach. His true position is extraordinarily complex, and amounts to a non-relationist position involving fundamentally spaceless objects known as monads. We will not delve into these issues, but, for a good place to start, the interested reader is directed to John Earman's article "Was Leibniz a Relationist?" (in P. A. French et al. (eds.) *Studies in Metaphysics, Volume 4*, University of Minnesota Press, 1979: pp. 263–276).

4 Readers wanting more detail here are advised to consult chapter 17 of Roger Penrose's *Road to Reality* (Alfred A. Knopf, 2004).

5 This strategy, of making what naively seems to be a relational property a monadic one (internal to the object possessing it) is known as *Sklar's Manoeuvre*. On the surface it sounds like a workable proposal, however, it has been rather controversial: see Brad

Skow's "Sklar's Maneuver" (*The British Journal for the Philosophy of Science* 58(4), 2007: 777–786) for a contrary voice.

6 For a clear-headed philosophical analysis of these issues and more, see Sklar's *Philosophy and the Foundations of Dynamics* (Cambridge University Press, 2013).

7 Martin Gardner describes a 'real world' version of this thought experiment in "The Ozma Problem and the Fall of Parity" (in J. Van Cleve and R. E. Frederick (eds.), *The Philosophy of Right and Left: Incongruent Counterparts and the Nature of Space*, Springer, 1991: pp. 76–77). The Ozma *project* was an early attempt to communicate with other planets – 'Ozma' refers to the ruler of Oz in the *Wizard of Oz*. The problem was to design a language that could operate across cosmic boundaries (which would, of course, be characterized by their own idiosyncratic conventions). One such attempt involved a method of transmitting pictures by using a kind of 'data matrix' method in which binary code is sent to indicate whether a cell is dark or light. One might send instructions on how to build a piece of technology for example – recall how in the movie *Contact* aliens sent *us* instructions for building a wormhole generator. The problem is: how do we transmit information about whether to use, e.g. left or right-handed screws? They might well print their matrix with the instructions entirely the wrong way around relative to the instructions we sent. A possible solution is provided by having them utilize universal laws of physics that violate mirror symmetry.

8 Carl Hoefer (2000) has reconstructed Kant's argument in terms of ascribing 'primitive identities' (i.e. brute, non-qualitative facts that allow for comparisons across counterfactual situations: different possible worlds) to the points of space. This is needed to make sense of performing a reflection on the 'lone hand' world in such a way that it would generate a new, different possibility – he takes primitive identities to be too great a price to pay given that, he argues, ultimately the relationalist can also explain any facts that need to be explained ("Kant's Hands and Earman's Pions: Chirality Arguments for Substantival Space," *International Studies in the Philosophy of Science* 14(3): 237–256).

9 I refer the reader wishing to have a more technical account of this problem to Oliver Pooley's "Handedness, Parity Violation, and the Reality of Space" (in K. Brading and E. Castellani (eds.), *Symmetries in Physics: Philosophical Reflections* (pp. 250–280). Cambridge University Press.).

10 We can classify these topological invariants by invoking winding numbers that count the number of times a loop (in this case a trajectory of one of our travellers) wraps around the rolled up dimension. So in the case of the torus (m, n) refers to m windings around the hole and n windings around the handle. No winding at all around either would simply be represented by $(0, 0)$. If you have a taste for interesting mathematics like this, then I urge you to read Richard Evan Schwartz's *Mostly Surfaces* (AMS, 2011).

11 This reshaping can be understood in a variety of ways. For example, we might think of the Now as advancing forward as part of a 'growing block,' so that the future is not yet fixed though the past is. Or we might think of this in reverse as a 'shrinking block' so that future is 'eaten away' by the ever-advancing present. Or, as presentists argue, we might deny reality to anything but the distinguished present moment. Even independently of special relativistic considerations, it has been argued that the notions of 'Now' and 'present' are anthropocentric, amounting to nothing more than "simultaneous with this utterance" – see, e.g. J. J. C. Smart, *Philosophy and Scientific Realism* (Routledge and Kegan Paul, 1963: p. 137). Special relativity provides a means

of extending this kind of reasoning, linking the Now to a frame of reference rather than anything specifically anthropcentric.

12 Nicholas Maxwell has argued that this conclusion (that the world's events are ontologically fixed) should lead us to *reject* special relativity because it conflicts with the 'probabilism' of quantum mechanics – "Are Probabilism and Special Relativity Incompatible?" (*Philosophy of Science* 52, 1985: 23–43). This would make a good paper to 'cut your critical teeth on' in the light of the discussion in this section and Chapter 7. David Albert also argued that Minkowski spacetime is a hard place to become (or "unfold") if you're a quantum mechanical state, though he invokes more aspects than the probabilistic evolution of Maxwell – "Special Relativity as an Open Question" (in H.-P. Breuer and F. Petruccione (eds.), *Relativistic Quantum Measurement and Decoherence*, Springer, 1999: pp 1–13).

13 Interestingly, since the (three-dimensional) shapes of ordinary objects involve spatial extension (parts separated by space), they too are relativized to frames of reference. It has been argued, therefore, that intrinsic shapes, if they are to exist, should be transfigured into four-dimensional, relativistically invariant properties – for more on this point, see Yuri Balahov's *Persistence and Spacetime* (Oxford University Press, 2010).

14 Mark Hinchliff has defended the view (called 'cone presentism') that identifies the present moment with the surface of the past light cone, so that any light signals that have reached a point constitute that event's present – "A Defense of Presentism in a Relativistic Setting" (*Philosophy of Science* 67, *Supplement*, 2000: S575–S586). This is very hard to swallow for all sorts of reasons. Firstly, it relativizes things to points, so that there are as many presents as spacetime points. Secondly, if we manage to capture light from the first photon created after the Big Bang, then that qualifies as present. Revisions are necessary, but this might be a step too far.

15 Stein prefers to call events that related outside of the light cone of a point "causally alien." This perhaps better captures the issues Stein has with Putnam's argument, and strikes me as more appropriate than any of the alternatives, but alas the terminology never took hold – "A Note on Time and Relativity Theory" (*The Journal of Philosophy* 67(9), 1970: 289–294).

16 For an excellent philosophical-historical discussion of this episode, see John Stachel's "The Hole Argument and Some Physical and Philosophical Implications" (*Living Reviews in Relativity* 17 (2014): http://relativity.livingreviews.org/Articles/lrr-2014-1/download/lrr-2014-1BW.pdf).

17 Diffeomorphisms are somewhat difficult to explain properly in an elementary manner, but fortunately we don't really need the details for this example. They are, more or less, transformations (isomorphisms) along the same lines as translations mapping one point or region to another (i.e. they are maps ϕ from the manifold to itself or to some other manifold, $\phi: \mathcal{M} \to \mathcal{M}'$), but that satisfy properties having to do with continuity. In the case of the hole argument we use the action of such maps on fields, so that they have the effect of *dragging*, e.g. the metric field from one point of the manifold onto another (this action is distinguished by an asterisk, ϕ^*). So given a field $\mathfrak{F}$ (with a physical interpretation), which might be defined at a point p, say, the diffeomorphism gives us $\phi^*\mathfrak{F}$ defined at another point q. The value of $\phi^*\mathfrak{F}$ at q is the *same* as the value of $\mathfrak{F}$ at p (because $q = \phi(p)$), but the value of $\phi^*\mathfrak{F}$ at p is *not* the same as the value of $\mathfrak{F}$ at p. This highlights the way in which the points p and q of the manifold play an important role in comparing the diffeomorphic fields: if the points are *real* then the field that we have dragged around is truly different in the two cases. If we

have some dynamical equations that cannot tell us which is realized in the world, for some complete specification of initial facts, then we will have a case of indeterminism.

18 A fuller statement of the view Hoefer calls "metric field substantivalism" can be found in his "The Metaphysics of Space-Time Substantivalism" (*Journal of Philosophy* 93(1), 1996: 5–27). In this paper he eliminates much of the metaphysical baggage that bloated earlier responses to the hole argument (specifically, the notion of 'primitive identity' for spacetime points, that we discuss below).

CHAPTER 5 FURTHER ADVENTURES IN SPACE AND TIME

1 Actually, Hermann von Helmholtz has the distinction of the creation of a discworld with two-dimensional beings confined to it (with no knowledge of higher dimensions): "On the Origin and Meaning of Geometrical Axioms" (in P. Pesic (ed.) *Beyond Geometry: Classic Papers from Riemann to Einstein*, Dover Publications, 2007: pp. 53–68). In any case, Poincaré himself uses beings confined to the interior of a sphere (see next note), but it has become 'conventional' to speak of Poincaré's disk!

2 John Norton expresses this very clearly by noting that the observational consequences O follow from the conjunction of a geometry G *and* some physical theory P about the bodies traversing the geometry. That is: $G + P = O$. Of course, we can preserve O by tweaking either the geometry or the physical theories so long as we perform a compensatory adjustment on the other – this example can be found in Norton's exceptionally clear guide "Philosophy of Space and Time" (in M. Salmon (ed.), *Introduction to the Philosophy of Science*, Prentice-Hall, 1992: pp. 179–232).

3 For Poincaré, all we have to go on are the *observed motions of objects*. From these observations we make inferences to a spatial reality underlying this. But Poincare's response was that it is the group of possible transformations of objects that matters: this is invariant in the cases since the objects are observed to move in the same way in the two scenarios (the whole point of the example being that the same body of evidence is compatible with two conflicting visions of an underlying spatial reality). This is closely related to Felix Klein's Erlangen Programme in which spatial geometry is characterized by its group of motions. The motions are initially derived from our visual and tactual-motor experience of the world, in bringing about displacements and alterations of objects. The convention of Euclidean space is selected, according to Poincaré, precisely because its group of transformations is the closest match to the coarse (physical) group of displacements we experience in our encounters with the world. It is fascinating to see how what we consider to be 'pure' subjects of mathematics like group theory originate in such observations – for more on these origins, see P. Pesic's collection of the original papers: *Beyond Geometry: Classic Papers from Riemann to Einstein*, Dover Publications, 2007.

4 It is a fun exercise to try and come up with a counterexample that would lead one to definitively tell whether one lived on the surface of a sphere (such as the Earth: though only its surface, with no access to higher dimensions) or not. If you manage this feat, drop a line to the 'Flat Earth Society': http://www.tfes.org.

5 There are a number of famous cases in which what were thought to be conventional choices were no such thing. For example, David Malament demonstrated that simultaneity in special relativity (a standard example wheeled out by conventionalists) can be shown to be non-conventional (and can be uniquely defined) given certain

undeniable assumptions – see his "Causal Theories of Time and the Conventionality of Simultaneity" (*Noûs* 11, 1977: 293–300).

6 I prefer to distinguish such dualities from the standard conventionalist cases. For the reasons why, and a general overview of dualities, see, e.g. my "A Philosopher Looks at String Dualities" (*Studies in History and Philosophy of Modern Physics* 42(1): 54–67).

7 Newton was no slouch, and identified the basis of the problem in his *Principia*:

> In astronomy, absolute time is distinguished from relative time by the equation of common time. For natural days, which are commonly considered equal for the purpose of measuring time, are actually unequal. Astronomers correct this inequality in order to measure celestial motions on the basis of a truer time. It is possible that there is no uniform motion by which time may have an exact measure. All motions can be accelerated and retarded, but the flow of absolute time cannot be changed. The duration or perseverance of the existence of things is the same, whether their motions are rapid or slow or null; accordingly, duration is rightly distinguished from its sensible measures and is gathered from them by means of an astronomical equation. Moreover, the need for using this equation in determining when phenomena occur is proved by experience with a pendulum clock and also by ellipses of the satellites of Jupiter. ([34], p. 410)

Poincaré simply disagrees that duration is distinct from the various relative measures of duration: the measures are not 'measures *of* some real underlying quantity see Harvey Brown's *Physical Relativity* (Oxford University Press, 2005, §2.2.3) for more on this, including, in later chapters, the story followed into general relativity.

8 Hans Reichenbach expresses this point (that the metric of time, or duration, is a conventional element) nicely, as follows: "It is impossible in an absolute sense to compare two consecutive units of a clock if we nonetheless wish to call them equal, this assertion has the nature of a definition" ("Methods of Physical Knowledge" [1929]; reprinted in H. Reichenbach et al. (eds.) *Hans Reichenbach: Selected Writings 1909–1953, Volume Two*, Springer, 1978: p. 184). To establish sameness of duration requires what Reichenbach (and the logical empiricists) call a "coordinative definition": it is defined by definitional lineage to some observable phenomenon (yet *not* by experience itself); but as Reichenbach goes onto argue (similarly to Poincaré) any such coordinations (e.g. with the Earth's rotation, with atoms, with light rays, and so on) involve arbitrary elements (such as a notion of simultaneity, which is a spatial notion that suffers similarly from its own 'problem of congruence').

9 An excellent semi-popular treatment of atomic clocks can be found in Tony Jones' *Splitting the Second: The Story of Atomic Time* (IOP Publishing, 2000). A more advanced, though still very readable treatment of modern time measurement (including discussions of some of the issues raised here) is Claude Audin and Bernard Guinot's *The Measurement of Time: Time, Frequency and the Atomic Clock* (Cambridge University Press, 2001).

10 Note that while Newton's laws of motion do not by themselves imply absolute space (since the laws are the same in all uniformly moving frames): we are unable to determine whether events separated in time are spatially coincident – this is, of course, just the content of Galilean relativity. But temporal relationships between spatially separated events have a different status: here we *can* say whether two spatially distant events are simultaneous or not (according to Newton's theory). According to

Sklar [46], Newton was aware of this difference, which is why he utilizes *practical* (physical) arguments from astronomy to argue for the reality of absolute time, but thought experiments (the bucket and the globes arguments) to argue for absolute space and motion.

11 However, Eran Tal has made a good start in exposing many interesting features of time standards. See e.g. his "Making Time: A Study in the Epistemology of Measurement" (*The British Journal for the Philosophy of Science*, forthcoming) for a philosophical investigation of time standardization.

12 This does not mean that there *is* only one branch realized. In 'branching time' models the world literally (i.e. topologically) takes multiple courses, so that it is the tree that is realized, rather than a single branch – for a discussion of branching in relation to indeterminism, see (though note that it is rather logic-heavy): T. Placek, N. Belnap, and K. Kishida's "On Topological Issues of Indeterminism" (*Erkenntnis* 79, 2014: 403–436).

13 Of course, it does not help us much by defining propensities in terms of dispositions, since they are just as slippery! The basic idea is best explained by simply thinking of propensities as brute chancy features in the world. Karl Popper famously based such a view on radioactive decay (half life), which seemed to be an irreducibly chancy business – see his "The Propensity Interpretation of Probability" (*The British Journal for the Philosophy of Science* 10(37), 1959: 25–42).

14 Pérez Laraudogoitia offers an excellent summary of a range of supertasks, including more that are relevant to the issue of determinism, in his online encyclopaedia article: http://plato.stanford.edu/entries/spacetime-supertasks/.

15 The definition of a singularity in general relativity is much broader than this, and the association with infinite curvature is rather outmoded (though it does capture much of the *physical* interpretation). The more mathematical treatment involves the idea that a singularity is a kind of 'boundary' on which the curves in the spacetime (that might represent motions of observers) end. (For a technical account of singularities relevant to the concerns of this section, see Robert Geroch's "What is a Singularity in General Relativity?" *Annals of Physics* 48,1968: 526–540.)

16 There is far more to the story than this. A notable addition is provided by the notion of a 'naked singularity,' which is a singularity not clothed in the usual event horizon blocking any undetermined surprises from view. But without such a horizon to mask the goings on it is possible for something like the space invaders mentioned earlier to appear! It is a serious piece of physics to try and find ways to forbid such naked singularities (cosmic censorship hypotheses) from finding a home in our world. I refer the interested reader to the brilliant, though technically demanding, *Bangs, Crunches, Whimpers, and Shrieks: Singularities and Acausalities in Relativistic Spacetimes*, by John Earman (Oxford University Press, 1995). This includes a discussion of supertasks in a range of generally relativistic spacetimes (especially so-called 'Malament-Hogarth' spacetimes containing both infinite and finite length worldlines) that appear to be exploitable to test what appear to be unprovable mathematical conjectures that would require infinite time to complete (e.g. Goldbach's conjecture that every even number is the sum of two primes). The observer with the infinite worldline could simply crank through all even numbers testing whether the conjecture holds while the observer with the finite length worldline sits in waiting for the result to be relayed to them. Whether or not these are just quirky mathematical games or point to something deep about computability in the world remains a matter of debate.

CHAPTER 6 LINKING MICRO TO MACRO

1 Though the association of entropy with disorder is problematic it provides a useful mental foothold from which one can view the more precise combinatorial landscape. Physicist Arieh Ben-Naim has gone to great lengths to correct various misunderstandings of entropy (including the disorder interpretation!). See his book *Entropy and the Second Law: Interpretation and Misss-Interpretations* (World Scientific, 2012) for a very useful treatment. For a more elementary exposition, see his *Discover Entropy and the Second Law of Thermodynamics: A Playful Way of Discovering a Law of Nature* (World Scientific, 2010).

CHAPTER 7 QUANTUM PHILOSOPHY

1 It is rather curious that we don't construct quantum theories from the ground up, but always in this parasitic way using the well-established formal frameworks for classical theories (Hamiltonian and Lagrangian mechanics) as a host (see Alexei Grinbaum's "Reconstruction of Quantum Theory," *British Journal for the Philosophy of Science* 58(3), 2007: 387–408) – this is not to say that there are classical analogs for all things quantum, of course: the spin observable is distinctively quantum since it is related to Planck's constant. Some brave souls have tried to build a quantum OS from the ground up (known as 'reconstructing quantum theory'), rather than porting. The idea here is to find a set of 'principles' (or axioms) from which one can derive the framework of quantum mechanics, in much the same way that Einstein deduced the Lorentz transformations from the principles of constant light velocity and of relativity. These are very interesting, and they point toward a deeper understanding of quantum mechanics, but in this book we stick with the more orthodox approaches and issues. Neither, in this chapter, will we be much concerned with the porting process itself, but instead with the OS and the ported applications. However, there are interesting philosophical issues associated with quantization (porting), especially concerning its role as an 'inter-theory relation' stitching together quantum and classical, and also as a case study for looking at taking *limits* in physical theories (see A. Bokulich's *Reexamining the Quantum-Classical Relation*, Cambridge University Press, 2008).

2 See, for example, Lee Rozema et al. "Violation of Heisenberg's Measurement-Disturbance Relationship by Weak Measurements" (*Physical Review Letters* 109, 2012: 100404).

3 The Stern–Gerlach apparatus, in the cases that interest us, performs much the same function as a half-silvered mirror, splitting a population of particles into two: in this case, spin up and spin down. There are also similar curiosities to the double slit experiment. For example, if we were to initially split a beam into spin-up and spin-down, and then feed only those that are spin-down into another Stern–Gerlach machine that has its field oriented at right angles to the first (so that it measures spin-left and spin-right), then we would find a 50/50 mixture of lefties and righties. But the odd thing is that if we then feed just the lefties into the original apparatus (measuring up/down again), we also find a 50/50 splitting! However, if we recombine the lefties and righties, and then feed it into the original, we find that all the particles are spin-up.

4 You can see John Bell explaining the theorem that bears his name in a video from a lecture in 1990: http://cds.cern.ch/record/1049544.

5 For more details, see §6 of Healey's entry on "Holism and Nonseparability in Physics": http://plato.stanford.edu/entries/physics-holism.

CHAPTER 8 ON THE EDGE: A SNAPSHOT OF ADVANCED TOPICS

1 Recall also that the twins paradox, discussed in §4.3, provides a kind of *weak* time travel, in which you can travel into the future by traveling at high speeds, thereby ageing less than the 'external time' back home. It is time travel in the sense that your journey will take fewer years than those poor homebound folk will experience. Done correctly, you could 'travel' far into Earth's future by simply completing a roundtrip at a high enough speed for long enough.

2 Usually this will involve unphysical matter-energy types, such as negative energy (i.e. anti-gravity) or infinitely large objects. Chris Smeenk and Chris Wüthrich refer to this method of making universes "designer spacetimes," which gives the right idea (see their "Time Travel and Time Machines" in C. Callender ed., *The Oxford Handbook of Philosophy of Time*, Oxford University Press, 2011: p. 378).

3 There are a variety of largely technical results that exhibit such paradox-avoidant behavior due to Kip Thorne and others looking at classical billiard ball models in the presence of wormholes – see e.g. Fernando Echeverria, Gunnar Klinkhammer, and Kip Thorne's "Billiard Balls in Wormhole Spacetimes with Closed Timelike Curves: Classical Theory," *Physical Review* D 44(4), 1991: 1077–1099.

4 The related Fermi paradox concerns the probable existence of other more advanced civilizations: why haven't we seen *them* yet, he asks? The universe should be teeming with them. See philosopher Nick Bostrom discuss this paradox on *Closer to Truth*: http://www.closertotruth.com/series/where-are-all-those-aliens#video-3988.

5 A Turing machine is simply an idealized (abstract) 'device' that completes some task by running an algorithm in discrete steps that will take the machine through a series of states, with each subsequent state depending on the current state, some transition rules (essentially, the computer's program functioning as 'evolutionary laws'), and a particular symbol that is being read by the device – an infinite tape functions as a 'memory' register for the machine that can be read from and written to. The key idea is that this defines a notion of 'computability': a function is computable just in case there exists a set of instructions (an algorithm) that will result in the Turing machine completing the computation (thereby halting), given its infinite tape and infinite time. A *Universal* Turing machine is then a machine that can do any tasks that any other Turing machine can do – it is perhaps easiest to understand by viewing the abstract Turing machine as a software programme that runs on the universal Turing machine (the hardware). For more detail, see David Barker-Plummer's "Turing Machines": http://plato.stanford.edu/entries/turing-machine/.

6 Oxford physicist David Deutsch introduced this idea as the following physical principle: "Every physical system can be perfectly simulated by a universal model computing machine operating by finite means" ("'Law Without Law' in Physics," *Foundations of Physics* 16(6): p. 589) – Christopher Timpson argues against Deutsch's claim in "Quantum Computers: The Church–Turing Hypothesis Versus the Turing Principle" (in C. Teuscher (ed.) *Alan Turing: Life and Legacy of a Great Thinker*, Springer, 2004: pp. 213–240).

7 Computational complexity theorists divide problems into '(time) complexity classes' according to how fast the algorithm can be solved for a given input length – those that can't be solved (for which no algorithm exists) are 'undecidable.' Polynomial time simply means that given an input string of length n the time to compute it for some algorithm is given as a polynominal in n (e.g. $n^3 + 3n$). It is safe to think of this

as being a 'doable' (or 'feasible') problem in an ordinary sense. The link between this notion of efficient computability and being doable in polynomial time is known as 'Cobham's Thesis.' Exponential time problems are not doable given that the n figures as an exponent. (A further interesting aspect of time complexity is the idea of an $\mathcal{NP}$ (non-deterministic polynomial) problem. This can be 'checked,' rather than *solved*, in polynomial time if an 'oracle' gives you a clue [or given a random search process: hence "non-deterministic"]: a million dollar prize (and mathematical immortality) awaits the person that can prove whether these [checking versus solving] really belong in the same complexity-class or not – for the official problem-statement from the Clay Mathematics Institute funding the prize, see: http://www.claymath.org/sites/default/files/pvsnp.pdf.)

8 The speed-up comes from two sources: firstly, from the parallelism of superpositions (which allows one to 'store' 2^n numbers compared to some single number n in the classical case) and are such that the operators of quantum mechanics will act on all components at once; and secondly, the interference between branches of such superpositions (i.e. entanglement). This opens the door to using constructive and destructive interference (with a well-chosen operator) to reduce the number of steps needed to compute algorithms. The former parallelism might be approximated by a classical device, but the entanglement is a distinctly quantum affair. The full details of how this operates in the case of the Shor algorithm are too complex to go into here, but for a good discussion, see Chapter 15 of Christopher Moore and Stephan Mertens' *The Nature of Computation* (Oxford University Press, 2011).

9 Andrew Steane attempts to demolish the idea that quantum computers demand a many-worlds view in his "A Quantum Computer Only Needs one Universe" (*Studies in History and Philosophy of Modern Physics* 34(3), 2003: 469–478) – note, however, that Steane bases his views on the so-called "Holevo bound" limiting the amount of (roughly, human readable) classical information that can be transmitted in a quantum channel; something a little distinct from computation itself, in the sense we are assuming.

10 In fact, P. D. Welch (2008) proved that the upper limit on computational power appears as a universal constant in the relevant world – "The Extent of Computation in Malament–Hogarth Spacetimes," *British Journal for the Philosophy of Science* 59: 659.

11 For a good review, see Christian Wüthrich's "A quantum-information-theoretic complement to a general-relativistic implementation of a beyond-Turing computer" (*Synthese* 192(7): 1989-2008) – here he suggests that quantum gravity might ultimately hold the key to unraveling the physical form of the Church–Turing thesis.

12 If these concepts aren't familiar, consult chapter 15 of the second volume of Richard Feynman's *Lectures on Physics* (Addison Wesley, 1971).

13 David Wallace gives a very readable explanation of these issues in his "Time-Dependent Symmetries: The Link Between Gauge Symmetries and Indeterminism" (in K. Brading and E. Castellani (eds.) *Symmetries in Physics: Philosophical Reflections*, Cambridge University Press, 2003: p. 163). Note that one key reason for dealing with local symmetries is to enforce the specially relativistic requirement of 'local action' (that is, outlawing action-at-a-distance): what transformations are performed at a point x should not be of concern to a point y spacelike separated from x.

14 There is a sense in which this can be given a topological explanation. The presence of the solenoid effectively removes a region of space so that the path traced by the beam of electrons (surrounding the solenoid) cannot be shrunk to a point – or, in

other words, the path going through one slit cannot be deformed into the path going through the other slit, so that the space is non-simply connected. In the classical case the motion of the particles (a charge, q) is fully determined by the electric and magnetic fields only (via the Lorentz force law $F = q(\mathbf{E} + v \times \mathbf{B})$), so we don't face this kind of feature. For more on this approach, see Antigone Nounou's "A fourth way to the Aharonov–Bohm effect" (in K. Brading and E. Castellani (eds.) *Symmetries in Physics: Philosophical Reflections*, 2003: 174–200).

15 This kind of nonlocality (viewed as a kind of 'non-separability' along the same lines as quantum entanglement) is discussed in Richard Healey's, "Gauge theory and Holisms" (*Studies in the History and Philosophy of Modern Physics* 35(4), 2004: 619–642).

16 Note that distinct gauge theories might recommend distinct interpretations. For example, whether there is a gravitational version of the Aharonov–Bohm effect that will allow us to run the same 'virtues' and 'vices' options is not yet settled: see R. Chiao et al., "A Gravitational Aharonov–Bohm Effect, and Its Connection to Parametric Oscillators and Gravitational Radiation" (in D. C. Struppa and J. M. Tollaksen (eds.) *Quantum Theory: A Two-Time Success Story*, Springer, 2014, pp. 213–246).

17 The physicist Rudolf Haag takes the centrality of particle-*detections* as the basis for his own interpretation based on *events*. Objects as continuants (things that continue to exist over dense intervals of time: that persist) are largely inferential on this account: they constitute what he calls "causal ties" linking the discrete events that form the real data of our experience and that breathe life into the quantum world. Theory, according to Haag, simply consists in finding models for predicting the properties of such events (in the future) from the properties of other events (in the past). In this sense, fields are inferences just as much as particles – see his "Fundamental Irreversibility and the Concept of Events" (*Communications in Mathematical Physics* 132, 1990: 245–251) and also his "Quantum Theory and the Division of the World" (*Mind and Matter* 2(2), 2004: 53–66).

18 Christopher Timpson and David Wallace have tried to counter some of these objections with a view they label 'spacetime state realism' – see their paper "Quantum Mechanics on Spacetime I: Spacetime State Realism" (*British Journal for the Philosophy of Science* 61, 2010: 697–727).

19 If you are interested in digging deeper, Meinard Kuhlmann's very readable entry on philosophical aspects of quantum field theory would make a good starting point: http://plato.stanford.edu/entries/quantum-field-theory/.

20 For a philosophical discussion of zero-point energy, see Simon Saunders' "Is the Zero-Point Energy Real?" (in M. Kuhlmann, H. Lyre, and A. Wayne (eds.) *Ontological Aspects of Quantum Field Theory*, World Scientific: pp. 313–343). Also worth a look is Rugh and Zinkernagel's "The Quantum Vacuum and the Cosmological Constant Problem" (*Studies in History and Philosophy of Modern Physics* 33(4), 2001: 663–705) – this paper also contains a nice historical discussion of the cosmological constant problem (that is, the conflict between the general relativistic and quantum field theoretic calculations of the vacuum energy).

21 A thorough treatment of the first fifty years of quantum gravity research, during which emerged the 'special features' of general relativity (that cause problems for quantization), can be found in D. Rickles, *Covered in Deep Mist: The Development of Quantum Gravity, 1916–1956* (Oxford University Press, forthcoming).

22 Virtual particles are field quanta that are allowed to 'live' within some process

(such as a transition from one quantum state to another, as one finds in particle accelerators) thanks to the time-energy uncertainty relations: $\Delta E \cdot \Delta t \geq \hbar$. So long as energy is conserved at the level of the process as a whole, the internal goings on can seemingly violate energy conservation (in a way satisfying the relations). Such 'intermediate processes' are not directly observable, however. But they are an essential part of the calculation of the probability of going from some input state (e.g. particles in an input beam to an accelerator) to those coming out (after a collision). The virtual processes, involving virtual particles, have to be summed over, with each *possible* way of going from input to output playing a role. For a philosophical examination of virtual particles, see Tobias Fox's "Haunted by the Spectre of Virtual Particles: A Philosophical Reconsideration" (*Journal for General Philosophy of Science* 39(1), 2008: 35–51).

23 A brief outline of the various approaches to quantum gravity, along with their particular quirks and conceptual problems, can be found in D. Rickles and S. Weinstein's "Quantum Gravity": http://plato.stanford.edu/entries/quantum-gravity/.

24 The usual quantum field theories are all *local* (yes, it's an overused word. . .) in the sense that the field interactions occur at individual points of spacetime. Making the interactions local allows for the peaceful coexistence of special relativity and quantum theory, thus preserving causality (ruling out action-at-a-distance), as mentioned above. But, this locality (involving the stacking up of field interactions at the same spacetime point) leads to singularities, which in turn lead to divergences. The 'renormalization programme' led to a finite version of quantum field theory for electromagnetic interactions, known as renormalized QED. The path to this theory involves the introduction of a cutoff so that wavelengths (respectively energies) shorter (respectively higher) than the cutoff are ignored (making the divergent terms finite). Depending on the cutoff chosen, we get different values predicted for the physical quantities. In order to get rid of this unwanted dependence on what is only an arbitrary cutoff, and get the theory predictively back on track, a readjustment called renormalization is performed. The parameter values of this modified theory are inserted from their experimentally observed ('clothed') values – clothed, that is, by the swarm of virtual particles that are produced in line with the time-energy uncertainty relations. However, the theory and the parameters at this stage are dependent on the cutoff scale, which, again, is arbitrary, so we need to take the continuum limit, letting it go to zero (or the momentum go to infinity). When we do this we get *renormalized QED*, with quantities that are independent of the cutoff. Broadly speaking, if one needs to redefine only a finite number of parameters to absorb the infinities then the theory is renormalizable. Otherwise it is non-renormalizable. This used to be considered fatal, but nowadays a more pragmatic interpretation is adopted according to which a theory is 'sensitive' to scale, so that as the energy is varied the theory may or may not continue to be useful (applicable). This domain dependent approach is known as *effective field theory*: a theory will be effective only within a certain range of energies. The reason that theories can be effective in a certain range is that they are 'insensitive' to what is going on at other scales (especially at higher energies = shorter distances) and so are rendered relatively free from 'interference' from elsewhere, depending on only a few global properties of the higher energies (smaller distances). It is possible that quantum gravity is an effective field theory in just this sense. (Note that renormalization has received a fair amount of interest from philosophers: see, for example, Huggett and Weingard's "The Renormalization Group and Effective Field

Theories" (*Synthese* 102(1), 1995: 171–194); or, for an exceptionally clear, elementary guide to the way renormalization is actually done, see John Baez's entry at: http://math.ucr.edu/home/baez/renormalization.html.

25 The best introduction to this problem, in my view, is to be found in Gordon Belot and John Earman's paper "Presocratic Quantum Gravity" (in C. Callender and N. Huggett (eds.) *Physics Meets Philosophy at the Planck Scale*, Cambridge University Press, 2001: 213–255).

26 Here possible worlds are viewed as being just as real as the actual world (our world). Possibility and necessity are then reduced to the space of such worlds by thinking of possibility as *truth at a world* and necessity as *truth at all worlds*. So, 'possibly *X*' simply means that there is a world at which *X*. See David Lewis' book *On the Plurality of Worlds* (Blackwell, 1986) for the canonical exposition.

27 Earman, "Thoroughly Modern McTaggart: Or, What McTaggart Would Have Said if He Had Read the General Theory of Relativity" (*Philosophers' Imprint* 2(3), 2002: 1–28) and Maudlin, "Throughly Muddled McTaggart: Or how to abuse Gauge Freedom to Generate Metaphysical Monstrosities" (*Philosophers' Imprint* 2(4), 2002: 1–19).

28 There is a fairly rich history of interactions between physics and the human sciences. Economics especially has a close connection, and a great many of its concepts and tools are taken directly from physics (and, in some cases vice versa). The classic treatment of this history is Philip Mirowski's *More Heat than Light: Economics as Social Physics, Physics as Nature's Economics* (Cambridge University Press, 1989). A more recent account that I can heartily recommend is James Owen Weatherall's *The Physics of Wall Street* (Mariner Books, 2014).

29 The neologism "econophysics" was coined by the statistical physicist (of critical phenomena) H. E. Stanley: see Stanley et al., "Anomalous Fluctuations in the Dynamics of Complex Systems: From DNA and Physiology to Econophysics" (*Physica A* 224, 1996: 302–321).

30 Benoit Mandelbrot has identified various examples of such fractal, scale-invariant financial data – see his *The (Mis)behavior of Markets: A Fractal View of Financial Turbulence* (Basic Books, 2006).

31 In his paper, "Explaining Financial Markets in Terms of Complex Systems" (*Philosophy of Science* 81(5), 2014: 1117–1130), Meinard Kuhlmann has argued that "Phase transitions in ferromagnets and financial markets can be studied in a common framework because the same structural mechanisms can be invoked in both cases" (p. 1222). This, he argues, renders econophysics "explanatorily fruitful." However, a mechanical account usually relies on the existence of lawlike behavior integral to the mechanism; but the non-stationarity might seem to cause problems in this respect: the spins in a magnet will have their statistical properties fixed for all time, but not so for the statistical properties of financial systems. However, the area known as 'renormalization group theory' shows that there are universal properties in such many-body systems (systems in the same universality class), meaning that diverse systems share the same critical exponents (and scaling behavior) and so display qualitatively identical macroscopic properties (when approaching criticality), for a certain class of 'fluctuation-based' properties. This might therefore be a useful tool with which to defend Kuhlmann's idea.

32 Leonard Susskind, who coined this name, discusses the nature of this ensemble, and the role of the Anthropic principle in string theory, in his book *The Cosmic Landscape: String Theory and the Illusion of Intelligent Design* (Back Bay Books, 2006).

33 For a defence of this point see John Earman's "The SAP also Rises: A Critical Examination of the Anthropic Principle" (*Philosophical Quarterly* 24(4), 1987: p. 316).

34 James Hartle and Mark Srednicki consider this ambiguity in the definition of observers, arguing that we don't have grounds for assuming that we are typical: "We have data that we exist in the Universe, but we have no evidence that we have been selected by some random process. We should not calculate as though we were" ("Are we Typical?" *Physical Review D* 75, 2007: 123523-1–123523-6). But every scrap of data should be included in our calculations.

35 Steven Weinstein ("Anthropic Reasoning and Typicality in Multiverse Cosmology and String Theory", *Classical and Quantum Gravity* 23, 2006: 4231–4236) argues that WAP should really be divided into two sub-types:

WAP$_1$: "What we can expect to observe must be restricted by the conditions necessary for our presence" (p. 4234).

WAP$_2$: "What we can expect to observe must be restricted by the conditions necessary for the presence of observers" (p. 4234).

Depending on which of these we choose there will be a large reference class in which we are not going to be typical. But if we choose *ourselves* as the reference class then we are trivially typical. However, it isn't completely clear that we know what "us" means in this case. For example, what if future humankind figures out how to 'upload' brains to silicon-based computing equipment: do we change our ideas of what a typical observer is? For a useful discussion of this point, see Feraz Azhar's "Prediction and Typicality in Multiverse Cosmology" (*Classical and Quantum Gravity* 31, 2014: 2–11).

References

1. Aaronson, S. (2013) *Quantum Computing since Democritus*. Cambridge University Press.
2. Barrett, M. and E. Sober (1992) Is Entropy Relevant to the Asymmetry Between Retrodiction and Prediction? *British Journal for the Philosophy of Science* **43**(2): 141–160.
3. Butterfield, J. (1988) Albert Einstein meets David Lewis. In A. Fine and J. Leplin (eds.), *PSA 1988, Volume 2*: 56–64.
4. Carter, B. (1974) Large number coincidences and the Anthropic principle in cosmology. In M. Longair (ed.), *Confrontation of Cosmological Theories with Observational Data* (pp. 291–298). D. Reidel.
5. Clifton, R. and M. Hogarth (1995) The Definability of Objective Becoming in Minkowski Spacetime. *Synthese* **103**(3): 355–387.
6. Deutsch, D. (1997) *The Fabric of Reality*. Penguin.
7. Dirac, P. A. M. (1958) *The Principles of Quantum Mechanics*. Clarendon Press.
8. Earman, J. and J. D. Norton (1987) What Price Spacetime Substantivalism? The Hole Story. *The British Journal for the Philosophy of Science* **38**: 515–525.
9. Earman, J. (1989) *World Enough and Space-Time: Absolute versus Relational Theories of Space and Time*. MIT Press.
10. Esfeld, M. (forthcoming) The Reality of Relations: The Case from Quantum Physics. In A. Marmodoro and D. Yates (eds.), *The Metaphysics of Relations*. Oxford University Press.
11. Esfeld, M. and V. Lam (2010) Ontic Structural Realism as a Metaphysics of Objects. In A. Bokulich and P. Bokulich (eds.), *Scientific Structuralism*. Springer.
12. Feynman, R. P. (1965) *Lectures on Physics, Volume III*. Addison-Weseley Publishing Company.
13. Feynman, R. P. (1981) Simulating Physics with Computers. *International Journal of Theoretical Physics* **21**(6–7): 467–488.
14. Feynman, R. P. (2006) *QED: The Strange Theory of Light and Matter*. Princeton University Press.
15. Frigg, R., S. Bradley, H. Du, and L. Smith (2014) Laplace's Demon and the Adventures of his Apprentices. *Philosophy of Science* **81**(1): 31–59.
16. Gambini, R. and J. Pullin (1996) *Loops, Knots, Gauge Theories, and Quantum Gravity*. Cambridge University Press.
17. Gardner, M. (1952) Nature Ambidextrous? *Philosophy and Phenomenological Research* **13**(2): 200–211.
18. Glymour, C. (1972) Topology, Cosmology, and Convention. *Synthese* **24**: 195–218.
19. Glymour, C. (1977) The Epistemology of Geometry. *Noûs* **1**: 227–251.
20. Gödel, K. (1949) A Remark About the Relationship Between Relativity Theory and Ideal. In P. Schilpp (ed.), *Albert Einstein: Philosopher-Scientist* (pp. 555–562). Open Court.

21. Grünbaum, A. (1973) *Philosophical Problems of Space and Time*. Reidel.
22. Hawking, S. W. (2001) Chronology Protection: Making the World Safe for Historians. In S. W. Hawking et al. (eds.), *The Future of Spacetime* (pp. 87–108). W.W. Norton.
23. Healey, R. (2006) Symmetry and the Scope of Scientific Realism. In W. Demopoulos and I. Pitowsky (eds.), *Physical Theory and its Interpretation* (pp. 143–160). The Western Ontario Series in Philosophy of Science. Springer.
24. Hoefer, C. (1998) Absolute versus Relational Spacetime: For Better or Worse, the Debate Goes On. *The British Journal for the Philosophy of Science* **49**(3): 451–467.
25. Huggett, N. (1994) What are Quanta, and Why Does it Matter? *PSA 1994, Vol. 2*: 69–76.
26. Huggett, N. (ed.) (2010) *Everywhere and Everywhen: Adventures in Physics and Philosophy*. Oxford University Press.
27. Ismael, J. and B. van Fraassen (2003) Symmetry as a Guide to Superfluous Theoretical Structure. In E. Castellani and K. Brading (eds.), *Symmetry in Physics: Philosophical Reflections* (pp. 371–392). Cambridge University Press.
28. Kant, I. (1768) Concerning the Ultimate Foundation of the Distinction of the Directions in Space. In B. Kerford and D. E. Walford (eds.), *Kant: Selected Pre-critical Writings and Correspondence* (pp. 36–43). Barnes and Noble, 1968.
29. Lewis, D. (1976) The Paradoxes of Time Travel. *American Philosophical Quarterly* **13**: 145–52.
30. Lucas, J. (1973) *A Treatise on Time and Space*. Methuen.
31. Luminet, J-P. (2011) Time, Topology, and the Twins Paradox. In C. Callender (ed.), *The Oxford Handbook of the Philosophy of Space* (pp. 528–545). Oxford University Press.
32. Maudlin, T. (1989) The Essence of Spacetime. In A. Fine and J. Leplin (eds.), *PSA 1988, Volume 2*: 82–91.
33. Nagel, E. *The Structure of Science: Problems in the Logic of Scientific Explanation*. Harcourt, Brace, and World.
34. Newton, I. (1999) *The Principia: Mathematical Principles of Natural Philosophy. A New Translation by I. Bernard Cohen and Anne Whitman*. University of California Press.
35. Pérez Laraudogoitia, J. (1996) A Beautiful Supertask. *Mind* **105**(417): 81–83.
36. Poincaré, H. (1898) On the Foundations of Geometry. In P. Pesic (ed.), *Beyond Geometry: Classic Papers from Riemann to Einstein* (pp. 117–146). Dover Publications, 2007.
37. Poincaré, H. (1913) *The Value of Science*. New York: The Science Press.
38. Putnam, H. (1967) Time and Physical Geometry. *Journal of Philosophy* **64**: 240–247.
39. Reichenbach, H. (1958) *The Philosophy of Space and Time*. Dover Publications Inc.
40. Richter, B. (2006) Theory in Particle Physics: Theological Speculation Versus Practical Knowledge. *Physics Today* **59**(10): 8–9.
41. Rynasiewicz, R. (1994) Absolute Versus Relational Space-Time: An Outmoded Debate?. *The Journal of Philosophy* **93**(6): 279–306.
42. Savitt, S. (2011) Time in the Special Theory of Relativity. In C. Callender (ed.), *The Oxford Handbook of Philosophy of Time* (pp. 546–570). Oxford University Press.
43. Schlosshauer, M., J. Kofler, and A. Zeilinger (2013) A Snapshot of Foundational Attitudes Toward Quantum Mechanics. arXiv: http://arxiv.org/pdf/1301.1069v1.pdf.
44. Schrödinger, E. (1935) Discussion of Probability Relations Between Separated Systems. *Proceedings of the Cambridge Philosophical Society* **31**: 555–563; **32** (1936): 446–451.

45. Schrödinger, E. (1950) What is an Elementary Particle? Reprinted in E. Castellani (ed.), *Interpreting Bodies: Classical and Quantum Objects in Modern Physics* (pp. 197–210). Princeton University Press, 1998.

46. Sklar, L. (1974) *Space, Time, and Spacetime*. University of California Press.

47. Sklar, L. (1990) Real Quantities and their Sensible Measures. In P. Bricker and R. I. G. Hughes (eds.), *Philosophical Perspectives on Newtonian Science* (pp. 57–76). MIT Press.

48. Stein, H. (1991) On Einstein–Minkowski Space-Time. *The Journal of Philosophy* 65(1): 5–23.

49. Stein, H. (1991) On Relativity Theory and the Openness of the Future. *Philosophy of Science* 58: 147–167.

50. Teller, P. (2001) The Ins and Outs of Counterfactual Switching. *Nous* 35(3): 365–393.

51. Ulam, S. (1991) *Adventures of a Mathematician*. University of California Press.

52. Van Fraassen, B. (1991) *Quantum Mechanics: An Empiricist View*. Oxford University Press.

53. Weyl, H. (1949) *Philosophy of Mathematics and Natural Science*. Princeton University Press.

54. Weyl, H. (1952) *Symmetry*. Princeton University Press.

55. Wheeler, J. A. (1990) Information, Physics, Quantum: The Search for Links. In W. H. Zureck (ed.), *Complexity, Entropy, and the Physics of Information* (pp. 309–336). Addison Wesley.

56. Wigner, E. (1964) Symmetry and Conservation Laws. In E. Wigner (ed.), *Symmetries and Reflections: Scientific Essays* (pp. 14–27). University of Indiana Press (reprinted, 1967).

Index

Page numbers in *italic* refer to figures and captions

Milton Keynes UK
Ingram Content Group UK Ltd.
UKHW021642160324
439538UK00007B/338

9 780745 669823